Physiology
Past, Present and Future

Yngve Zotterman and his friends

Bristol, July 1979

Left to right: Alex von Muralt, Richard Adrian, Lloyd Beidler, Carl Pfaffmann,
Yngve Zotterman, Manfred Zimmermann, George Gordon, Sven Landgren, Herbert Hensel,
Dan Kenshalo, Ainsley Iggo.

Physiology
Past, Present and Future

A Symposium in honour of
Yngve Zotterman

held in the

Department of Physiology, Medical School, University of Bristol
July 11th & 12th 1979

Edited by

D. J. ANDERSON
Professor of Oral Biology, University of Bristol

PERGAMON PRESS

OXFORD · NEW YORK · TORONTO · SYDNEY · PARIS · FRANKFURT

U.K.	Pergamon Press Ltd., Headington Hill Hall, Oxford OX3 0BW, England
U.S.A.	Pergamon Press Inc., Maxwell House, Fairview Park, Elmsford, New York 10523, U.S.A.
CANADA	Pergamon of Canada, Suite 104, 150 Consumers Road, Willowdale, Ontario M2J 1P9, Canada
AUSTRALIA	Pergamon Press (Aust.) Pty. Ltd., P.O. Box 544, Potts Point, N.S.W. 2011, Australia
FRANCE	Pergamon Press SARL, 24 rue des Ecoles, 75240 Paris, Cedex 05, France
FEDERAL REPUBLIC OF GERMANY	Pergamon Press GmbH, 6242 Kronberg-Taunus, Hammerweg 6, Federal Republic of Germany

First edition 1980

British Library Cataloguing in Publication Data
Physiology - past, present and future.
1. Neurophysiology - Addresses, essays, lectures
2. Nervous system - Mammals
- Addresses, essays, lectures
I. Anderson, Declan John
II. Zotterman, Yngve
599'.01'88   QP361   80-40957
ISBN 0 08 025480 2

In order to make this volume available as economically and as rapidly as possible the authors' typescripts have been reproduced in their original forms. This method has its typographical limitations but it is hoped that they in no way distract the reader.

Printed and bound in Great Britain by
William Clowes (Beccles) Limited, Beccles and London

CONTENTS

LIST OF PARTICIPANTS

S. Adams,
Bristol,
England.

Lord Adrian,
Cambridge,
England.

D.J. Anderson,
Bristol,
England.

K. Appenteng,
London,
England.

S. Barasi,
Cardiff,
Wales.

R.H. Baxendale,
Glasgow,
Scotland.

L. Beidler,
Florida,
U.S.A.

F.R. Bell,
Hatfield,
England.

K. Bhoola,
Bristol,
England.

T.J. Biscoe,
London,
England.

M.A. Bishop,
Bristol,
England.

D.R. Bowsher,
Liverpool,
England.

J.P. Boyle,
Glasgow,
Scotland.

S. Cadden,
Bristol,
England.

Sandra Carpenter,
London,
England.

R.M. Cash,
London,
England.

F.J. Clark,
Oxford,
England.

B.A. Cross,
Cambridge,
England.

W.R. Ferrell,
Glasgow,
Scotland.

O. Franzén,
Uppsala,
Sweden.

P. Gjørstrup,
Lund,
Sweden.

G. Gordon,
Oxford,
England.

Sir John Gray,
Plymouth,
England.

R.L. Gregory,
Bristol,
England.

Elizabeth Gregson,
Bristol,
England.

A. Guz,
London,
England.

P. Hayes,
Alverstoke,
England.

H. Hensel,
Marburg,
W. Germany

W. Hamann,
London,
England.

P. Harrison,
London,
England.

P.M. Headley,
Paris,
France.

R. Hill,
Bristol,
England.

R. Holland,
Winnipeg,
Canada.

A. Iggo,
Edinburgh,
Scotland.

C.A. Keele,
Radlett,
England.

D. Kenshalo,
Florida,
U.S.A.

List of Participants

S. Landgren,
Umeå,
Sweden.

R.W. Linden,
London,
England.

D. Lloyd,
Pulborough,
England.

S.D. Logan,
Birmingham,
England.

B.M. Lumb,
Birmingham,
England.

B. Lynn,
London,
England.

B. Matthews,
Bristol,
England.

R. Morris,
Bristol,
England.

P. Nathan,
London,
England.

S. Northeast,
Bristol,
England.

J. Ochoa,
Hanover,
U.S.A.

K. Olsson,
Umeå,
Sweden.

R. Orchardson,
Glasgow,
Scotland.

C. O'Reilly,
Belfast,
N. Ireland.

H. Patel,
Birmingham,
England.

Sharon Pay,
Cardiff,
Wales,

C. Pfaffmann,
New York,
U.S.A.

C.G. Phillips,
Oxford,
England.

T. Quilliam,
London,
England.

P.P. Robinson,
Bristol,
England.

D. Scott, Jr.,
Philadelphia,
U.S.A.

A. Taylor,
London,
England.

C.C.N. Vass,
Edgware,
England.

A. von Muralt,
Arniberg,
Switzerland.

G. Walsh,
Edinburgh,
Scotland.

J.H. Wolstencroft,
Birmingham,
England.

R. Yemm,
Dundee,
Scotland.

M. Zimmermann,
Heidelberg,
W. Germany.

Y. Zotterman,
Stockholm,
Sweden.

PREFACE

The idea for this Symposium was conceived following the Satellite Symposium entitled "Pain in the Trigeminal Region" in Bristol in 1977, after the I.U.P.S. Congress in Paris. As President of I.U.P.S. and Chairman of the I.U.P.S. Commission on Oral Physiology, Yngve Zotterman swept aside many of the early difficulties which faced us in organising that symposium and during the meeting he showed a breadth of knowledge and intellectual liveliness of which many people half a century his junior would be proud.

When I learned - to my surprise - that Yngve was to reach the age of 80 in September 1978, and recalling his enjoyment of the Bristol Symposium in 1977, I thought that he would welcome another meeting in Bristol to celebrate his great age and achievement, in the company of some of his distinguished friends and collaborators. It was neither possible nor desirable to have the meeting at the time of his 80th birthday when he was to be with his family and very close friends in Stockholm, so July 1979 was chosen, when we dared to hope that the weather might allow us to enjoy the occasion to the full. Not only was the weather kind to us, but so also were his friends who came to do him honour, the many organizations which provided financial and other material support, and my own colleagues who, in many large and small ways, made my task easy and the occasion successful.

It is hard to credit that Yngve Zotterman entered the Karolinska Medical School over 60 years ago in 1916 and that he joined an advanced class in physiology in Langley's laboratory at Cambridge in 1919 where he was taught by Joseph Barcroft, A.V. Hill, H. Hartridge and of course, E.D. Adrian, with whom he returned to work in 1920 and again in 1925. Yngve's first paper presented at the Physiological Society and published in the Proceedings in 1920, was entitled "The conductivity of the nerve ending in acid solutions". Five years later he went back to Cambridge and in 1926 with Adrian published two papers entitled "The impulses produced by sensory nerve endings" (Parts 2 and 3) on which his reputation as a neurophysiologist was to be built. It is perhaps not so well known that during this period in England he worked for a while with Sir Thomas Lewis and with Lewis published two papers on the vascular reactions of the skin. Towards the end of 1926 he joined A.V. Hill in experiments on heat production of nerve. His research has also ranged over heavy work in lumber-men and deep water diving and for many years since 1935 he has studied the neural mechanisms of taste. It was Yngve who made the first recordings in conscious human subjects of subjective sensations in response to gustatory stimuli, relating them to the nerve impulse pattern in the chorda tympani.

Yngve's energy seems limitless and the list of distinguished organizations on which he has served is too long for me to include. For five years from 1967-1972 he was President of the Trustees of the Nobel Foundation and from 1971-1974 he was President of the I.U.P.S. He is secretary of the Wenner Gren Centre in Stockholm and still organizes symposia at the Centre. In 1973 he founded the "Working group for the study of the problems of elderly people" and has edited 7 volumes of symposia already. I know that there are more to come.

ix

Yngve has received many tributes at home and abroad to his academic achievements, none I suspect more proudly accepted than the Sc.D. at the University of Cambridge in 1968, conferred on him by his old friend E.D. Adrian, who was Chancellor of the University at the time.

This is the man in whose honour we foregathered in Bristol in July 1979. Those of us who were there are glad to know that Yngve greatly enjoyed the occasion, as we did also.

Declan Anderson

ACKNOWLEDGEMENTS

Most grateful thanks are extended for financial assistance in running this Symposium to:

The Wellcome Trust	Parke Davis & Co.
Astra Chemicals	Pergamon Press
Beecham Products	Pharmacia (G.B.)
Cordent Dental Trust	Reckitt & Colman
Grass Instrument Co.	Stag Instruments
Kabivitrum	Carl Zeiss (Oberkochen)
Medelec	

The University of Bristol kindly provided hospitality.

INTRODUCTION

I find myself writing this introduction for what one might call family reasons. But birthdays are occasions when family feelings are important and I was indeed delighted to have been asked to be at the meeting which this volume commemorates. I was there as a naive observer to learn about new work in a field which I think I might be said to have grown up with - at any rate I grew up in a world bestrode by neurophysiological collosi of whom Yngve was one. I hardly thought when I first met Yngve that I should have the opportunity and pleasure of hearing him talk on the future of Physiology, with all his experience and the wisdom of 81 years. To be honest, I think I was hardly thinking of physiology at all - I daresay I wasn't even thinking.

Our first encounter, I'm told, was more than 50 years ago, so I can, I think, claim a lifelong friendship as an excuse for having been at the meeting. Over the years, my childhood affection has matured into scientific admiration. It was great fun to congratulate Yngve in his eighty-first year. We listened, as I think my father would have said, to lots of really exciting stuff and I know how very much pleasure it would have given him to have been at the meeting to add his congratulations to the birthday celebrations. Yngve was in at the very beginning of sensory electrophysiology and his contributions form part of what every biologist knows. Indeed, they know it almost before they are taught it, so much has it become part of general knowledge. He has continued over his entire life to be amongst those who with their own hands have added to scientific knowledge and I should like to make a prediction about the future, and that is that in the field of sensory neurophysiology, Yngve's fundamental work will continue for years to inform and inspire both our experimental work and our ideas.

Richard Adrian

THE SAGA OF THE 'C' FIBRES

A. Iggo

Department of Veterinary Physiology, University of Edinburgh

ABSTRACT

The non-myelinated axons in peripheral nerves were originally reported in 1838 by
Remak , although their detailed description had to await the introduction of the
electron microscope. Most of them are dorsal root afferent, but some enter the
spinal cord via the ventral roots. Their functional characteristics, and particularly the rôle in pain attracted attention in the 1930's. The introduction in
1956 of single unit C-fibre recording yielded exact information needed for a satisfactory analysis of the receptor characteristics. A broad survey is made of C-
fibre afferents in the viscera, skin and deeper somatic tissues. The action of
these C-fibres in the central nervous system, principally in the dorsal horn of
the spinal cord is briefly reviewed.

KEYWORDS

Afferent C-fibres, non-myelinated axons, cutaneous receptors, visceral receptors,
muscle receptors.

INTRODUCTION

The existence in peripheral nerves of very small axons was first demonstrated by
Remak in 1838, who microdissected peripheral nerves in water. He saw very fine
interlacing filaments with nucleated corpuscles, subsequently found to be Schwann
cell nuclei (Fig. 1A). These filaments have become known as Remak bundles or
Remak fibres. Tuckett (1895) using a technique of teasing the nerve apart in
aqueous humour, was able to distinguish three components in Remak's fibres - nuclei
of the sheath (Schwann) cells, a sheath and the cores of the fibres (Fig. 1B). The
C, or more correctly, the non-myelinated (non-medullated, unmyelinated) fibres in
peripheral nerve thus have a long scientific history. They were revealed with
improved clarity in Ranson's modification of Cajal's silver stain for axons. He
found them to arise principally from dorsal root ganglion cells (Ranson, 1911) and
to be segregated in the lateral division of the spinal dorsal roots in cats and to
enter the dorsal horn immediately, or after a short passage in Lissauer's tract.
This separation is clearly marked in primates. It gave an opportunity to assess
experimentally their sensory rôle. When this lateral division of the dorsal root
was sectioned by Ranson in cats, the 'pseudo-affective reflexes' were found to be
weaker in response to previously effective noxious stimuli. Although the dorsal

1

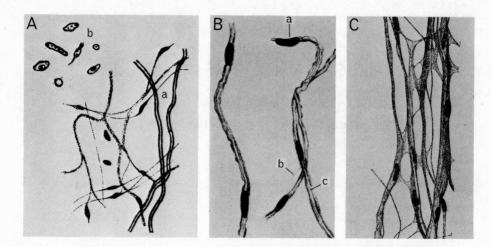

Fig. 1. Diagrams to illustrate successive stages in the recognition of
C-fibres in peripheral nerves. A. Remak's original (1838) drawing
of nerve fibres dissected in water. The thick large fibres (a)
are myelinated axons with an axis cylinder and the interlacing
filaments (fibrae organicae) are the Remak bundles of C-fibres.
The corpuscular bodies (b) are what we now call Schwann cell nuclei.
B. Tuckett's (1895) drawings of Remak's fibres, showing (a) the
nuclei of the sheath cells (b) sheath and (c) cores of the fibres,
equivalent to axis cylinders. C. Nageotte's (1922) figure of cat
cervical sympathetic C-fibres, showing clearly the anastomosing
system. Only the satellite (sheath) cells branch, the axons are
continuous unbranched individuals.

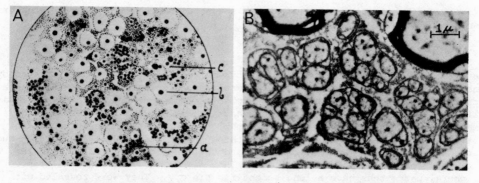

Fig. 2. Histological advances. A. Ranson's (1911) original illustrations
of pyridine silver stained peripheral nerve, showing a) non-myelinated fibres, b) large myelinated fibres and c) small myelinated
fibres. B. Gasser's (1955) electron micrograph of nerve showing a
myelinated axon and several bundles (Remak's fibres) of non-myelinated axons.

root afferents that run into Lissauer's tract include both small myelinated and
non-myelinated axons, the idea was established that the small fibres, including
C-fibres, were pain fibres. There are also, of course, numerous C-fibres which
are part of the autonomic nervous system.

This classical Bell-Magendie view of the afferent fibres has been modified by the
work of Coggeshall and Willis and their colleagues on the spinal ventral roots.
These roots have long been known to contain C-fibres and also neuronal cell bodies.
Coggeshall, Coulter and Willis (1974) have now established that many ventral root
C-fibres are afferent. The C-fibres comprise about thirty per cent of the ventral
root fibres and about half of them are sensory. There are therefore substantial
numbers of afferent C-fibres that enter the small cord through the ventral roots.
Only a very small number of myelinated afferent fibres take the same course
(Applebaum and others, 1976). Clifton and others (1976) dealing with the sacral
and caudal roots of the cat report that the afferents are distributed widely among
the pelvic viscera as well as to the skin and deeper somatic structures. In general
they had relatively insensitive receptors. Those in the viscera were either mucosal,
responding to severe mechanical stimuli, or were distension sensitive. The new
results present a challenge to the classical Bell-Magendie Law but little is yet
known of the sensory or reflex functions of the ventral root afferent C-fibres.
They could be nociceptors.

ELECTROPHYSIOLOGY. WAVES AND UNIT POTENTIALS

The introduction in 1924-6 by Adrian (1926) of techniques for electrical recording
of sensory impulses from nerve endings and particularly the results obtained by
Adrian and Zotterman (1926) using a single end organ preparation both heralded a
new epoch in neurophysiology and gave promise that the C-fibres would soon reveal
their secrets. Not so. In 1931 in his Croonian lecture Adrian was able to sur-
mise that something must be going on in the small fibres, but the capillary electrometer was too insensitive to reveal what it was. In fact, when we consider the kind
of evidence provided by the capillary electrometer used in the original studies and
Matthews' oscillograph used for the later studies (Fig. 3), and the constraints they
imposed on the investigators, we can only admire the success achieved.

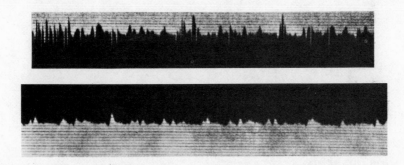

Fig. 3. Records obtained using the Matthews' oscillograph of action
        potentials recorded by Zotterman (1933) from dorsal cutan-
        eous nerves of the Frog. Upper record, response to pin
        thrust into the skin at start of tracing. Lower record,
        response to radiant heat applied to the skin, causing slow
        action potentials to appear in the record.

Now enters the C-wave of the compound action potential which has given its name to
the laboratory jargon for the non-medullated axons in peripheral nerves. Gasser
and Erlanger introduced and developed the application of the cathode-ray oscillo-
scope and thermionic amplifiers to the point where, in excised peripheral nerves,
it was possible to detect, first a synchronous volley of impulses in the largest
and fastest myelinated axons (the A wave), then the slower B wave and finally after
some doubts and refinement of the techniques, the C-wave (Erlanger and Gasser, 1930).
The correlation of these waves with the various sizes of axon in peripheral nerve
occupied Erlanger, Gasser and Bishop for several years but the main picture was
clear by the time of the publication of Erlanger and Gasser's monograph in 1937,and
the C-wave was established as originating in the non-myelinated axons. In 1955
Gasser returned to the C-fibres and showed in the sural nerve of the cat that they
conduct at velocities between 2.3 and 0.6 m/s, thus amending the original conduction
rates "of the C-fibres 1.6 - 0.3 m.p.s" (Erlanger and Gasser,1930).

Meantime, the search for the axons of receptors with high sensitivity to noxious
stimuli continued in Europe using the technique of unit recording.   An active
laboratory in this search was that of our Guest of Honour, Yngve Zotterman.   He
followed up the advances that flowed from the introduction of Matthews' sensitive
amplifying and recording system that replaced the capillary electrometer.   Even
with this new device Adrian, Cattell and Hoagland in 1931 were unable to subdue
the C-fibres. Meantime, the existence of large and small afferent fibres and their
relative sensitivity to conduction block by ischaemia, cold, local anaesthetic
and/or pressure had led Zotterman (1933) to give strong support to the view that
first and second (burning pain) were due to impulses in myelinated and non-myelinated fibres respectively and thus to promote the view that the C-fibres were pain
fibres. Attempts to separate the different waves of the compound action potential
in peripheral nerve established that the C-wave was more readily blocked by local
anaesthetic and less sensitive to ischaemia than the largest components of the
A-wave. This gave promise of an attribution of sensory functions to the C-fibres
in peripheral nerves. The A-fibres were already regarded as necessary for tactile
sensory inputs. Clarke, Hughes and Gasser in 1935 showed that the C-fibres
mediated pain but that the localized cutaneous pain was mediated by small myelinated
axones. In man, knowledge about differential block in peripheral nerve of A and C-
fibres, as well as the existence of first and second 'pain' led to the attribution

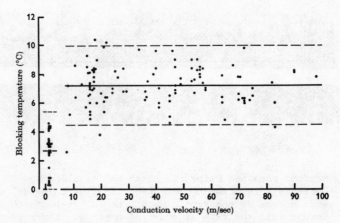

Fig. 4. Relation between conduction velocity of axons in the saphen-
        ous nerve and the temperature at which conduction is blocked.
        ● - myelinated axons, O - non-myelinated axons. The mean
        blocking temperatures were 7.2°C for the A-fibres and 2.7°C
        for the C-fibres (from Franz and Iggo, 1968).

of second pain to C-fibres by Zotterman in 1933. Even here the claims were contested, and recent single unit studies have shown that differential block by cold (Paintal, 1965; Franz and Iggo, 1968), local anaesthetic (Franz and Perry, 1974) and probably by ischaemia is less selective than once thought. Although A and C-fibres can be separated by selective block (Fig. 4) the various A-fibres cannot, particularly when natural repetitive stimulation via receptors is used as the input. This is due to the frequency-related Wedensky inhibitory actions (Paintal, 1965). This can even prevent a separation of A and C-fibres. Although the C-fibres can still conduct at temperatures below the absolute blocking temperature for A-fibres (Franz and Iggo, 1968) they may nevertheless block at higher temperatures, if the firing rate in the C-fibres is high.

The story continues, with an analysis by Zotterman (1936, 1939) of the effect of recording conditions, and particularly of the axon diameter and the correlated inter-electrode resistance on the signal:noise ratio and the spike amplitude of individual axons in unit recording. As a result he was able to secure from fine fascicles of peripheral nerves, records of the impulses in small mammalian axons. These were published in the Journal of Physiology in 1939 and the records clearly show very small waves of relatively long-time course that could be evoked by tactile and by noxious stimulation of the skin (Fig. 5). Because of the simultaneous presence in the records of larger action potentials and the uncertain number of axons generating the small waves, it was a matter of fine judgement to decide whether a particular fibre had a unique receptor characteristic. The small waves

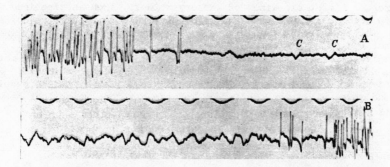

Fig. 5. Action potentials from a small fascicle of the saphenous nerve of a cat, in response to firmly stroking the skin, twice a second. The records read from right to left. In A, the start of the stroke was towards the left, setting up large impulses and B, shows the end of the stroke, which was followed by an after-discharge of small irregular waves. Individual small waves, labelled C, are present at the beginning of the upper record. These are residual after-discharges following a previous stroke. The fine marks are 50/s (Zotterman, 1939).

appeared in response to various natural stimuli, but in particular they could be evoked by holding a naked flame close to, or on, the skin. Here then were the long sought mammalian afferent units fitted to the task of signalling 'pain'. Their assignment to C-fibres by Zotterman was based on their spike-amplitudes using his analysis of the influence of recording conditions on spike size. The small size of the impulses, however, prevented conduction velocity measurement and Zotterman concludes that "As direct data of the velocity of these very small spikes is still lacking, it is possible that the potentials here recorded travel at lower rates than is indicated by their spike heights". (Zotterman, 1939).

C-fibre studies received their next impetus from the electronic developments hast-
ened by military conflict, which began to affect electrophysiology in the 1950's.
Gasser (1950, 1955) returned to the task of reconstructing the C-wave in the compound action potential of cutaneous nerves, greatly aided by the dramatic new
knowledge revealed by the electron microscope. He discovered that Remak's fibre
was a group of non-myelinated axons, supported by a Schwann cell and that the axons
maintained a physical separation from each other. He confirmed Nageotte's (1922)
discovery (Fig. 1C) that the sheaths of the C-fibres are arranged in anastomoses,
but that the axons are individual and non-anastomosing (Fig. 2B). Gasser (1955)
showed that there was no suprathreshold interaction between the axons which might
have resulted from their close apposition, and concluded that "Indeed, one can with
some assurance assume that they [the dorsal root C-fibres] conform to the general
rule of isolated conduction. He thereby raised the possibility that individual
C-fibres carry distinct and separable action potentials. His ultrastructural
results, however, raised the spectre that the small size and neighbourly habits of
the C-fibres would together frustrate any attempt to separate them by microdissection of the peripheral nerve.

The 1950's also saw a renewal of the earlier intensity of electro-physiological
studies of sensory mechanisms and the small axons of peripheral nerves presented a
challenge, although major interest was centred on the large myelinated axons
supplying muscle receptors. Maruhashi, Mizuguchi and Tasaki (1952) reported a
careful correlation of cutaneous axonal diameter and receptor function, and in the
toad were able to record from C-fibres in bundles also containing a few fine myelinated fibres. "Almost all kinds of sensory stimuli, such as pinching, pricking,
burning with acid or applying heat or cold, give rise to an outburst of afferent
impulses". They also found that light tactile stimuli were generally ineffective
and that "hot and cold endings" were insensitive to mechanical stimuli.   When
their microdissection technique was used on the cat their success was more limited,
but as Fig. 6 shows noxious and heat stimuli were effective excitants. Stroking
the skin set-up after-discharges in the unmyelinated fibres. These results thus
extended, and in some measure confirmed, Zotterman's conclusion that he was recording activity in C-fibres.

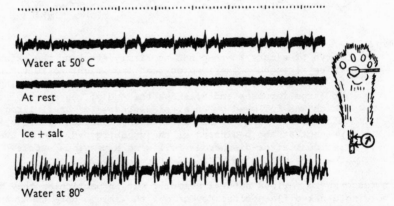

Water at 50° C

At rest

Ice + salt

Water at 80°

Fig. 6. Afferent impulses recorded from a small group of fibres incl-
        uding a number of unmyelinated fibres. The fibres were pre-
        pared by microdissection of the plantar nerve of the cat for
        recording by a modification of Tasaki's "bridge insulator"
        method (Maruhashi, Mizuguchi and Tasaki, 1952).

SINGLE UNIT ANALYSIS

Interest in visceral reflexes forced an attack on the small axons, since they form
the majority in visceral nerves, and resulted in the identification of a variety
of cardiovascular, respiratory and gastro-intestinal receptors with small myelinated axons. The prevailing uncertainty, even of the existence of impulses in
mammalian C-fibres encouraged caution. When unit action potentials in the cervical
vagus conducting at 2 m/s were reported (Iggo, 1955) they were attributed to myelinated axons, even though the C-waves of the compound action potential of the sural
and saphenous nerves have their fastest components travelling at 2.3 m/s. One
problem was the higher rate of conduction in the proximal part of some axons, so
that although the peripheral part of the axon was unmyelinated it had a myelin
sheath more proximally (Iggo, 1958). A recent detailed study by Duclaux, Mei and
Ranieri (1976) confirms the result. It became necessary to obtain more detailed
measurements of conduction velocity along the axon (Iggo, 1958) to support the
claim that impulses were recorded from individual C-fibres.

Improvements in technique led in 1956 to the first report (Iggo, 1956) of identified single unit action potentials in afferent fibres conducting at velocities well
down in the C-range (at 0.8 to 1.1 m/s) in the cervical vagus of the cat (Fig. 7).
No claim that these were C-fibres was made at the time, but in subsequent more
detailed publications (Iggo, 1958) this was asserted and supported by further
evidence. Not that it was immediately accepted, in part because the evidence was
principally electrophysiological. Confirmation that impulses could be recorded
from C-fibres was provided by Bower (1959, 1966) who recorded unitary action
potentials, conducting at 0.4 to 1.5 m/s in rabbit uterine nerve, which "certainly
came from non-myelinated fibres as they were yielded by nerves containing no myelinated fibres at all" (Bower, 1966). The stage was set for an advance into previously uncharted waters.

Additional techniques that reduced the need to make single fibre preparation and
thus minimize the demanding delicate microdissection of peripheral nerves are, for
example, the 'collision technique' which allows individual units in a multi-unit
strand to be identified (Iggo, 1958; Brown and Iggo, 1967). Nevertheless, in the
early experiments, the microdissection yielded nerve strands containing only a
single active axon in continuity. It may be asked how such a situation could arise
given the fact that the C-fibres run together in Remak bundles in peripheral nerves
(Gasser, 1955). The answer probably lies in Gasser's confirmation of Nageotte's
discovery (Fig. 1C) that there is a continual interchange of individual axons
between Schwann bundles. Gasser made serial reconstruction based on cross-sections
of a 210 µm length of saphenous nerve. Within that length no axons were parallel
for more than a few tens of microns. Since the dissection method depends on longitudinal separation of nerve strands into finer filaments the cross bridges between
adjacent Schwann bundles would be broken and therefore only very few (more often,
none!) of the axons in any bundle would remain in continuity with the distal main
nerve fascicle. This microdissection method continues to be the most productive
one in current use, judging from the number of papers published. Nevertheless, it
is tedious and other techniques have been introduced in an attempt to make the C-
fibres more accessible to study.

Douglas and Ritchie (1957a) offered one way out, with a technique aimed at sampling
the activity of a population of C-fibres in a peripheral nerve. This was achieved
by collision of orthodromic naturally induced impulses with a synchronous electrically induced antidromic volley (the "antidromic occlusion technique"). The reduction
in amplitude of the antidromic volley was used as an index of the degree of activation of the relevant axonal population. When applied to a nerve with a homogeneous
population of afferent units, within a particular fibre spectrum, the method yields

valuable results. The cutaneous C-fibres were reported by Douglas and Ritchie to
comprise one population with receptors sensitive to low intensity mechanical stimulation and to cooling and another, more slowly-conducting set, that was excited
by noxious mechanical stimulation, but not, rather surprisingly, by high skin temperatures (Douglas and Ritchie, 1957b; Douglas, Ritchie and Straub, 1960).

Another method uses microelectrodes to record from the dorsal root, or corresponding
visceral, ganglion cells. Mei (1962) introduced this method to record extracellul-
arly from the nodose ganglion and subsequently reported successful recording from
both myelinated and non-myelinated vagal afferent fibres. The majority of the C-
fibres supplied abdominal viscera. The same technique was later used by Bessou and
others (1971) with the added refinement of intracellular recording. In their hands,
C-ganglion cells (dorsal root somata) when penetrated with 3 M KCl glass micropipettes gave large resting and action potentials. "in most cases, the action
potential declined" as the electrode remained in the cell and more stable results
(from a few seconds to two hours) were obtained by extracellular recording.

Microelectrodes have also been inserted directly into peripheral nerves. The glass
micropipettes record satisfactorily from the larger myelinated axons whereas
insulated tungsten microelectrodes have been found to record in addition from non-
myelinated axons. The percutaneous recording of peripheral axonal spikes in man,
introduced by Vallbo and Hagbarth (1968), uses electrodes with a tip diameter of
1 to 5 μm. The method has now been used in several laboratories and has yielded
excellent single unit records of afferent and sympathetic C-fibres (Hallin and
Torebjörk, 1974; van Hees and Gybels, 1972) although the difficulties of the method
should not be underestimated.

From the foregoing brief review it is clear that the C-fibres in peripheral mammalian nerves are now fully available for unit analysis. We can be confident that,
with proper attention to detail, a comprehensive picture of the unit responses of
both afferent and efferent fibres can be built-up. It need hardly be said that in
parallel with the progress made in C-fibre unit analysis, there has been a progressive sophistication in instrumentation, with quantitative control of various stimulus
parameters, improved data-capture and computational aids to data analysis.

Where have we got to ?   And perhaps of special interest to this Symposium, how
good were the original conclusions when viewed in the light of our present knowledge
I shall proceed by considering in order the visceral, cutaneous and muscle afferent
C-fibre systems, but shall do no more than make a general survey since space and
time permits no more.

                    VISCERAL AFFERENT C-FIBRES

Most of the axons in visceral nerve are unmyelinated and the majority of the C-fibre
are afferent (see Douglas and Ritchie, 1962 for a review). Several kinds of afferen
unit have been found in single unit studies.

In-Series Tension Receptors

The original identified C-fibre single unit recordings were made in the cat (Iggo,
1956, 1957a, b) while examining mechano-sensitive gastric and intestinal receptors
(Fig. 7). These continue to be the most commonly reported gastro-intestinal
receptors, with afferent fibres in either the vagi or the splanchnic nerves. The
characteristic response is a discharge during distension of the viscus and during
contraction of that part of the wall in which the receptor lies. The receptors are
slowly-adapting and sustain a continuous discharge of impulses during steady
distension, so long as the smooth muscle does not relax. With prolonged distension,

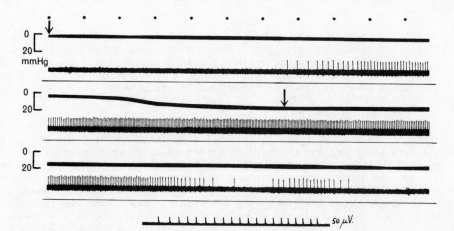

Fig. 7. Single unit recording of a vagal afferent C-fibre, conduction velocity 1 m/s. The receptor was in the pyloric antrum and the discharge of impulses was elicited by distension of the whole intact stomach. Inflow of fluid began at the arrow in upper trace, and ended at the arrow in second trace. The fluid remained in the stomach for the remainder of the record. The records are continuous. Time marks at 1 s intervals (from Iggo, 1957a).

accompanied by smooth muscle relaxation, the discharge becomes intermittent, with burst of impulses, that can be correlated with visible movement of the smooth muscle (Iggo, 1957a). Subsequent studies in various species (cat, rat, sheep, rabbit) have confirmed the general characteristics of this class of alimentary canal receptor (Morrison, 1977).

Splanchnic Mechanoreceptors

These were reported by Gernandt and Zotterman (1946) to be excited by intestinal movements. Morrison (1977) reports that many C-fibres in the splanchnic nerves can be excited by localized mechanical stimulation with a slowly-adapting response. A particularly effective stimulus was traction on the roots of the mesenteric blood vessels. In contrast to the 'in series' tension receptors with vagal afferent fibres, these splanchnic mechanoreceptors a) had up to 9 scattered receptor sites and b) were not consistently excited by either contraction or distension of the viscera. In addition a wide range of sizes of axon were involved, including both Aδ and C-fibres.

Mucosal receptors with C-afferent fibres include those originally described as pH receptors (Iggo, 1957b) on the basis of their responses to solutions of various pH. Although this classification has been questioned, it is still not clear quite what the effective stimulus is for this kind of receptor, at least for those in the glandular mucosa of the stomach and intestines (Davison, 1972). A new addition to the presumed intestinal C-afferent receptors is the 'glucoreceptor' of Mei (1978). These units are excited by intraluminal perfusion of the intestine with D-glucose (1 - 20 g/l) and give a slow-adapting response. Although the exact location of the receptors was not established, Mei argues that they are in the mucosa. Stimuli that activate the pH receptors and 'in series' tension receptors were without effect, nor were they affected by osmotic changes. In this respect they are like the 'sweet'

receptors of the sheep's tongue (Iggo and Leek, 1967), although the latter had
myelinated axons.

These new results that continue to be reported are, perhaps, less surprising than
might be thought, given the very large numbers (thousands) of non-myelinated
afferent fibres in the visceral nerves and the quite small total sample of single
C-units reported in the literature. The existence of a substantial body of new
information about visceral afferent mechanisms is undoubted.

CUTANEOUS C-FIBRES

The majority of the non-myelinated afferent fibres innervate the skin and so there
has been a persistent curiosity about their functional characteristics. In 1931
Adrian in his Croonian lecture concluded that the response to injury of the skin
might be conducted along C-fibres, although electrophysiological experiments at
that time had failed to detect the unitary impulses. In the ensuing decade there
were various attempts, as for example by Zotterman in his electrophysiological
unit recording studies and by others using reflexes (Clark, Hughes and Gasser,1935)
or differential nerve block experiments in man to establish the nature of the nociceptor afferents. This emphasis tended to concentrate attention on a putative
nociceptive rôle for the C-fibres.

Although Zotterman (1939) attributed a variety of effective stimuli to cutaneous
C-fibres, including light stroking, needle prick and burning, it was not possible
to assert categorically what the various kinds of unit comprised. He states that
the "C" spikes in his records came from "fibres below 5.5 µm diameter" and that
"some C potentials at least seem to be unmyelinated fibres". Problems associated
with the small size of the spikes and the usually irregular shape of the potentials
prevented the satisfactory use of his technique for measuring conduction velocities.

In 1957 Douglas and Ritchie using their whole nerve "antidromic occlusion technique"
reported that many afferent C-fibres in skin nerves could be excited by innocuous
mechanical stimuli delivered to the skin, as also could cooling the skin. Other
C-fibres were excited by vigorous mechanical stimuli, but none were detected that
responded to a very typical painful stimulus - heating the skin. The results gave
a new dimension to the cutaneous C-fibres. When the single unit recording technique
developed for visceral nerves was used on skin nerves (Iggo, 1958 et seq.;
Iriuchijima and Zotterman, 1960) the predictions of Zotterman (1939) and the general
conclusions of Maruhashi, Mizuguchi and Tasaki, (1952); Douglas and Ritchie (1957b)
were quickly confirmed. The cutaneous C-fibres could be compared and contrasted
with the A-fibre receptors that were also under concurrent unit scrutiny.

The situation in the skin reveals the particular strength and need for a single
unit analysis. Because there are several kinds of C-afferent supplying the skin
our knowledge of their specific characteristics remains inexact until each can be
examined in isolation from the remainder. It turns out, as might be expected,
that the different kinds of unit have only a relative, not absolute specificity.
The use of severe, even extreme, stimulation has particular risks for the C-fibres.
In the search for responses to noxious stimulation, it is not unlikely that receptors highly sensitive to other forms of stimuli, may be excited by the noxious
stimulus being tested.

In summary, there are three major categories of skin C-receptors - mechanoreceptors,
thermoreceptors and nociceptors. These basic conclusions of the early single unit
work (Iggo, 1959, 1960; Hensel, Iggo and Witt, 1960) have been sustained by numerous
subsequent studies, that have added finer detail to the early single unit reports.
A concise summary of present knowledge can take the following form, with some
implication for sensory function.

a) <u>Mechanoreceptors</u>. Excited by movement of hairs or light indentation of the skin.
The terminals end in a small patch of skin (less than 3 x 2 mm) although more remote
stimuli, by stimulus spread, can excite the endings.   The mechanical sensitivity
varies from those excited by moving hairs to firm pressure on the skin.   The C-
mechanoreceptors form a coherent group, when quantitative stimulus/response characteristics are measured.   Their capacity to function quantitatively is, however,
severely limited by the inexcitability that is caused by repeated (Fig. 8) or
maintained mechanical stimulation (Iggo and Kornhuber, 1977).   The after discharge
that was a frequently noted feature of the earlier reports, is unlikely to be a
special property of the receptors, that makes them continue to respond long after
a stimulus is over.   Instead, it appears to be a consequence of the ability of the
receptor to respond to bi-directional and slow movements of the skin (Iggo and
Kornhuber, 1977).   After-discharge is prominent when the skin is stroked and in

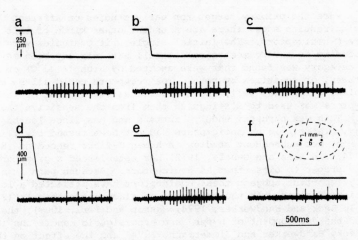

Fig. 8. C-mechanoreceptor in hairy skin of the cat. The receptive
field is shown as an inset in f. The receptor was indented
by 250 µm displacements by a fine-tipped probe at positions
a, b and c in the receptive field, with the responses shown
in the upper row of records. There is the characteristic
latency, between the onset of response and the first
recorded impulse, due to the slow conduction along the axon.
Next the outer points were stimulated twelve times at 4-
second intervals, and an inexcitability of the receptor
developed as indicated by the weak responses to larger
indentations (400 µm) in records d and f. The interjacent
locus b, has an almost undiminished sensitivity, indicating
that the inexcitability is a localized response, restricted
to the activated terminals. (from Iggo and Kornhuber, 1977).

these conditions slow restorative movement of the skin are visible on microscopic
examination. Bessou and others (1971) have also drawn attention to this characteristic feature of C-mechanoreceptors of responding very well to low-frequency
mechanical stimulation. Single unit analysis has established that individual C-
mechanoreceptors are excited by a fall in skin temperature but they are much less
sensitive than the C-cold receptors and Hahn (1971) reports that the mechanisms
yielding mechanical and thermal responses are moderately, but not completely
independent.

b) <u>Thermoreceptors</u>. The early reports of C-fibres responding to changes of skin temperatures were soon borne out by single unit studies. In fact the majority of both cold and warm receptors in hairy skin on non-primate mammals turn out, for the most part, to have non-myelinated axons. Hensel, Iggo and Witt (1960) made the initial detailed stimulus/response studies and found two clear categories of cold and warm fibres. The cold receptors had characteristics very similar to the lingual and cold receptors in the cat (Hensel and Zotterman (1951). The receptive fields were all spot-like, and the sensitivity of the receptors was comparable to human threshold detection. In primates many of the cold fibres have small myelinated axons whereas all the warm fibre axons are unmyelinated. In addition some of the cold fibres are as well, although the greater ease of investigating the unitary responses of the myelinated cold units has concentrated attention on them. There is no doubt, however, about the existence of sensitive warm and cold receptors with afferent C-fibres.

c) <u>Nociceptors</u> were the primary target for early studies on afferent C-fibres but, as the foregoing results show, there are abundant other kinds of unit. Nevertheless there are C-nociceptors. The initial single unit C studies were made on cats. C-fibres responding to severe mechanical stimuli have already been mentioned. A second major category was found that were excited by high (> 43°C) or low (< 20°C) skin temperatures, or sometimes by both. The receptors could all be excited by severe mechanical stimuli as well, and the original description of them as heat and cold receptors was used to distinguish them from the sensitive thermoreceptors (Iggo, 1959). This was perhaps, unduly cautious and has since led to a confusion of terminology. Their rôle as nociceptors had not been tested fully in the original description, but subsequent studies in human C-fibre recording (Hallin and Torebjörk, 1974; van Hees and Gybels, 1972) has established a good correlation, so that it is now proper to class them as nociceptors. Because pain is a powerful and clinically important subject these nociceptors have attracted a large number of investigations dealing with the responses to algogenic chemicals (Fjällbrant and Iggo, 1961; Beck and Handwerker, 1974; Bessou and Perl, 1969), their thermal thresholds and the development of hyper- and hypo-algesic conditions (Bessou and Perl, 1969;/ Beck, Handwerker and Zimmermann, 1974) and the effect on them of non-steroidal anti-inflammatory agents acting possibly via inhibition of prostaglandin synthetase (Vane, 1971). This considerable expansion of single unit studies is testimony both to the potential importance of the results and the difficulty of the subject matter.

DEEPER SOMATIC C-FIBRES

Interest in afferent C-fibres supplying the muscles has not been directed at their rôle in normal postural reflex activity since the large afferents to the muscle spindles and tendon organs are the prime receptor agents. Instead, there has been the assumption that the C-fibres were largely concerned with nocifensor functions. This was borne out by the early reports - by Bessou and Laporte (1960) using the "antidromic occlusion technique" and Iggo (1961) with single fibre recording. The C-fibres all had small receptive fields in muscle or tendon, were insensitive to muscle stretch, responding best to direct pressure with a small probe (Fig. 9). They were nearly all excited by both high and low temperatures. Classical deep-pain excitants, such as muscle ischaemia, excited only a low frequency discharge (Iggo, 1960). These basic features have been confirmed by subsequent work (Kumazawa and Mizumura, 1976), which in Schmidt's laboratory has also taken up the responses to algogenic and ergogenic chemicals (Mense and Schmidt, 1974). It is tempting to postulate these muscle receptors as the archetypal nociceptor were it not for the common fate of extinction that awaits such broad generalisations.

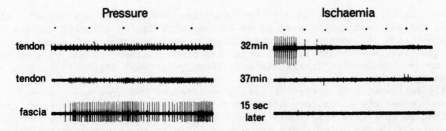

Fig. 9. Muscle C-nociceptors. The left-hand tracings show the response of
3 different C-fibres, in the same nerve strand, to direct mechanical stimulation of their receptors, in gastrocnemius muscle of the
cat. The right-hand tracings show the effect on the same units of
prolonged muscle ischaemia, coupled with direct electrical stimulation of the muscle to generate muscle contraction. The muscle had
become unresponsive before these latter records were made. There
is only a relatively weak excitation of the 3 afferent units with
large action potentials which disappear 15 s after restoring the
blood flow. (from Iggo, 1961).

CENTRAL ACTIONS OF THE C-FIBRES

All the above-mentioned study of C-fibres have as one end-product, a better understanding of the afferent input to the central nervous system. During the last two
decades there has been a virtual explosion of studies on somatosensory and visceral
sensory processing in the central nervous system. It is clearly out of the question
to attempt even the briefest review in this article. One feature of recent studies
has been the use of natural stimuli as inputs to the C.N.S. in laboratory investigation, in place of the almost exclusive use of electrical excitation of peripheral
nerve trunks that was a common feature of earlier studies. The account I have just
given of the diversity of receptors innervated by C-fibres goes some way to account-
ing for this recent trend. No electrical stimulus can select a particular category
of receptor, or afferent unit.

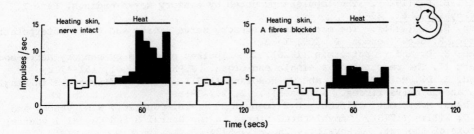

Fig. 10.Use of 'cold block' of myelinated fibres in a peripheral nerve, to
test the involvement of C-nociceptors in the activation of dorsal
horn neurones. In both histograms, the discharge of a Class 3 noci-
ceptor-driven neurone in lamina I of the lumbar dorsal horn is
plotted. There is a continuous background activity that is enhanced
when the skin is heated above 45°C, as shown by the blackened columns.
The response is still present when all the myelinated afferent fibres
in the sural nerve had been blocked, in the right figure. The C-
nociceptors are thus established as having an excitatory action
(unpublished record of Guilbaud, Iggo and Ramsay).

Detailed analytical studies impose their own demands on the investigators. For example, radiant heat can be used to avoid contamination of afferent input with additional kinds of receptors, and thus confuse the electrophysiological analysis. Such a quantitatively controlled stimulus has greatly aided the analysis of nociceptor and non-nociceptor interactions in sensory pathways (Handwerker, Iggo and Zimmermann, 1974). Although single unit studies have established that most of the thermally activated nociceptors in cats have afferent C-fibres there is still the question whether the excitatory afferent drive is carried by such afferent units. One method to test this hypothesis is to combine differential block of the peripheral nerve with the spinal cord experiment. Figure 10 illustrates the results obtained in such an experiment. A nociceptor driven cell is recorded in lamina I of the spinal cord. This is a Christensen and Perl (1970) lamina I unit, or a Class 3 neurone according to Iggo (1974), excited by noxious stimulation of the hairy skin, supplied by the sural nerve. That nerve was prepared according to Franz and Iggo (1968), so that the afferent input could include or exclude myelinated fibres. It is evident that the C-fibres carry a large part of the excitatory afferent drive elicited by heating the skin to noxious levels. By such a combination of techniques it should be possible to establish the relative importance of an input in various kinds of C-fibre. One puzzle is the rôle of the sensitive C-mechanoreceptors. Although the histological evidence is that the C-fibres terminate in the dorsal horn close to the root of entry, no neurones excited by light mechanical stimuli could be found in the experiments of the kind illustrated in Fig. 10.

CONCLUDING REMARKS

The saga of the C-fibres looks like becoming endless, as new facts in apparently endless succession are woven into the story. I would hope that our Guest of Honour is suitably rewarded for his early struggles by the subsequent knowledge that has been inspired by his studies. I am personally delighted to have had an opportunity to share in this present Symposium, and to indicate how I see the present state of our knowledge of the elusive C-fibres.

REFERENCES

Adrian, E. D. (1926). The impulses produced by sensory nerve endings. Part I. J.Physiol., 61, 49-72.

Adrian, E. D. (1931). The messages in sensory nerve fibres and their interpetation. Proc. R. Soc. (London) B, 109, 1-18.

Adrian, E. D. and Y. Zotterman (1926). The impulses produced by sensory nerve-endings Part II. The response of a single end-organ. J.Physiol., 61, 151-171.

Adrian, E. D., McK. Cattell, and H. Hoagland (1931). Sensory discharges in single cutaneous nerve fibres. J.Physiol., 72, 377-391.

Applebaum, M.L., G.L. Clifton., R.E.Coggeshall., J.D. Coulter., W.H. Vance, and W.D. Willis (1976). Unmyelinated fibres in the Sacral 3 and Caudal 1 ventral roots of the cat. J.Physiol., 256, 557-572.

Beck, P.W, and H.O. Handwerker (1974). Bradykinin and Serotonin effects on various types of cutaneous nerve fibres. Pflügers Arch. 347, 209-222.

Beck, P.W., H.O. Handwerker, and M. Zimmermann (1974). Nervous outflow from the cat's foot during noxious radiant heat stimulation. Brain Res. 67, 373-386.

Bessou, P., and Y. Laporte (1960). Activation des fibres afférentes myélinisées de petit calibre, d'origine musculaire (fibres du groupe III). C. R. Soc. Biol. 154, 1093.

Bessou, P., and E.R. Perl (1969). Response of cutaneous sensory units with unmyelinated fibers to noxious stimuli. J.Neurophysiol., 32, 1025-1043.

Bessou, P. P.R. Burgess., E.R. Perl, and C.B. Taylor (1971). Dynamic properties of mechanoreceptors with unmyelinated (C) fibers. J.Neurophysiol., 34, 116-131.

Bower, E.A. (1959). Action potentials from uterine sensory nerves. J.Physiol.,
    148, 2-3P.
Bower, E.A. (1966). The characteristics of spontaneous and evoked action potentials
    recorded from the rabbit's uterine nerves. J.Physiol., 183, 730-747.
Brown, A.G., and A. Iggo (1967). A quantitative study of cutaneous receptors and
    afferent fibres in cat and rabbit. J.Physiol., 193, 707-733.
Christensen, B.N, and E.R. Perl (1970). Spinal neurons specifically excited by
    noxious or thermal stimuli : marginal zone of the dorsal horn. J.Neurophysiol.,
    33, 293-307.
Clark, D., J. Hughes., and H.S. Gasser (1935). Afferent function in the group of
    nerve fibres of slowest conduction velocity. Am. J. Physiol. 114, 69-76.
Clifton, G.L., R.E. Coggeshall., W.H. Vance, and W.D. Willis (1976). Receptive
    fields of unmyelinated ventral root afferent fibres in the cat. J.Physiol.
    256, 573-600.
Coggeshall, R.E., J.D. Coulter, and W.D. Willis.Jr. (1974). Unmyelinated axons
    in the ventral roots of the cat lumbosacral enlargements. J.comp. Neurol.
    153, 39-58.
Davison, J.S. (1972). Response of single vagal afferent fibres to mechanical
    and chemical stimulation of the gastric and duodenal mucosa in cats.
    Q. Jl. exp. Physiology, 57, 405-416.
Douglas, W.W.,and J.M. Ritchie (1957). A technique for recording functional
    activity in specific groups of medullated and non-medullated fibres in whole
    nerve trunks. J.Physiol.,,138, 19-30.
Douglas, W.W.,and J.M. Ritchie (1957). Non-medullated fibres in the saphenous
    nerve which signal touch. J.Physiol., 139, 385-399.
Douglas, W.W.,and J.M. Ritchie (1962). Mammalian non-myelinated nerve fibers.
    Physiol. Rev. 42, 297-334.
Douglas, W.W.,J.M. Ritchie, and R.W. Straub (1960). The role of non-myelinated
    fibres in signalling cooling of the skin. J.Physiol., 150, 266-283.
Duclaux, R., N. Mei, and F. Ranieri (1976). Conduction velocity along the afferent
    vagal dendrites : a new type of fibre. J.Physiol. 260, 487-495.
Erlanger, J.,and H.S. Gasser (1930). The action potentials in fibers of slow
    conduction in spinal roots and somatic nerves. Am.J.Physiol., 92, 43-82.
Erlanger, J.,and H.S. Gasser (1937). The electrical signs of nervous activity.
    University of Pennsylvania Press, Philadelphia.
Fjällbrant, N.,and A. Iggo (1961). The effect of histamine, 5-hydroxytryptamine
    and acetylcholine on cutaneous afferent fibres. J.Physiol., 156, 578-590.
Franz, D.N.,and A. Iggo (1968). Conduction failure in myelinated and non-myelinated
    axons at low temperatures. J.Physiol., 199, 319-345.
Franz, D.N., and R.S. Perry (1974). Mechanisms for differential block among single
    myelinated and non-myelinated axons by procaine. J.Physiol., 236, 193-210.
Gasser, H.S. (1950). Unmedullated fibers originating in dorsal root ganglia.
    J. gen. Physiol., 33, 651-690.
Gasser, H.S. (1955). Properties of dorsal root unmedullated fibers on the two
    sides of the ganglion. J. gen. Physiol., 38, 709-728.
Gernandt, B., and Y. Zotterman (1946). Intestinal pain : an electrophysiological
    investigation on mesenteric nerves. Acta. physiol. scand., 12, 56-72.
Hahn, J.F. (1971). Thermal-mechanical stimulus interactions in low-threshold C-
    fiber mechanoreceptors of cat. Expl. Neurol., 33, 607-617.
Hallin, R.G., and H.E. Torebjörk (1974a). Single unit sympathetic activity in
    human skin nerves during rest and various manoeuvres. Acta.physiol.scand.,92,
    303-317.
Hallin, R.G., and H.E. Torebjörk (1974b). Methods to differentiate electrically
    induced afferent and sympathetic C-unit responses in human cutaneous nerves.
    Acta. physiol.scand., 92, 318-331.

van Hees, J., and J. Gybels (1972). Pain related to single afferent C fibers from
    human skin. Brain Res., 48, 397-400.

Hensel, H., and Y. Zotterman (1951). Quantitative Beziehungen zwischen der Entladu einzelner Kältefasern und der Temperatur. Acata. physiol. scand., 23, 291-319.

Hensel, H., A. Iggo, and I. Witt (1960). A quantitative study of sensitive cutaneous thermoreceptors with C afferent fibres. J.Physiol., 153, 113-126.

Iggo, A. (1955). Tension receptors in the stomach and the urinary bladder. J.Physiol. 128, 593-607.

Iggo, A. (1956). Afferent fibres from the viscera. XXth. Int.physiol.Congr. pp.458-459.

Iggo, A. (1957a). Gastro-intestinal tension receptors with unmyelinated afferent fibres in the vagus of the cat. Q.Jl.exp.Physiol., 42, 130-143.

Iggo, A. (1957b). Gastric mucosal chemoreceptors with vagal afferent fibres in the cat. Q.Jl. exp.Physiol. 42, 398-409.

Iggo, A. (1958). The electrophysiological identification of single nerve fibres, with particular reference to the slowest conducting vagal afferent fibres in the cat. J.Physiol., 142, 110-126.

Iggo, A. (1959). Cutaneous heat and cold receptors with slowly-conducting (C) afferent fibres. Q.Jl.exp.Physiol., 44, 362-370.

Iggo, A. (1960). Cutaneous mechanoreceptors with afferent C fibres. J.Physiol., 152, 337-353.

Iggo, A. (1961). Non-myelinated afferent fibres from mammalian skeletal muscle. J.Physiol., 155, 52-53P.

Iggo, A. (1974). Activation of cutaneous nociceptors and their actions on dorsal horn neurons. In J.J. Bonica (Ed.). Advances in Neorology, Vol. 4, Raven Press New York. pp. 1-9.

Iggo, A., and H.H. Kornhuber (1977). A quantitative study of C-mechanoreceptors in hairy skin of the cat. J.Physiol. 271, 549-565.

Iggo, A., and B.F. Leek (1967). The afferent innervation of the tongue of the sheep. In Y. Zotterman (Ed.). Proceedings of the 2nd.International Symposium Olfaction and Taste II. Pergamon Press, Oxford. pp. 493-507.

Iriuchijima, J., and Y. Zotterman (1960). The specificity of afferent cutaneous C fibres in mammals. Acta.physiol. scand., 49, 267-278.

Kumazawa, T., and K. Mizumura (1976). The polymodal C-fiber receptor in the muscle of the dog. Brain Res., 101, 589-593.

Maruhashi, J., K. Mizuguchi, and I. Tasaki (1952). Action currents in single afferent nerve fibres elicited by stimulation of the skin of the toad and the cat. J.Physiol., 117, 129-151.

Mei, N. (1962). Enregistrement de l'activité unitaire des afférences vagales. Réception par microélectrodes au niveau du ganglion plexiforme. Annls.Biol. anim.Biochim.Biophys. 2, 361-364.

Mei, N. (1978). Vagal glucoreceptors in the small intestine of the cat. J.Physiol. 282, 485-506.

Mense, S., and R.F. Schmidt (1974). Activation of group IV afferent units from muscle by algesic agents. Brain Res., 72, 305-310.

Morrison, J.F.G. (1977). The afferent innervation of the gastrointestinal tract. In Frank P. Brooks and Patricia W. Evers (Eds.) Nerves and Gut. Charles B. Slack Inc. Thorofare, N.J. pp. 297-326.

Nageotte, J. (1922). L'Organisation de la Matiere dans ses Rapports avac la Vie. Alcan, Paris.

Paintal, A.S. (1965) Block of conduction in mammalian myelinated nerve fibres by low temperatures. J.Physiol., 180, 1-19.

Ranson, S.W. (1911). Non-medullated nerve fibers in the spinal nerves. Amer.J.Anat. 12, 67-87.

Remak, R. (1838). Observationes anatomicae et microscopicae de systematis-nervosi structura. Berlin.

Tuckettt, I.L. (1895). On the structure and degeneration of non-medullated nerve fibres. J.Physiol., 19, 267-311.

Vallbo, A.B., and K.-E. Hagbarth (1968). Activity from skin mechanoreceptors
   recorded percutaneously in awake human subjects. Expl. Neurol., 21, 270-289.
Vane, J.R. (1971). Inhibition of prostaglandin synthesis as a mechanism of action
   for aspirin-like drugs. Nature New Biol., 231, 232-235.
Zotterman, Y. (1933). Studies in peripheral nervous mechanism of pain.
   Acta med. scand. 80, 9-64.
Zotterman, Y. (1936). Specific action potentials in the lingual nerve of the cat.
   Scand. Arch. Physiol.,75, 105-119.
Zotterman, Y. (1939). Touch, pain and tickling : an electrophysiological invest-
   igation on cutaneous sensory nerves. J.Physiol., 95, 1-28.

TOUCH AND PAIN—FACTS AND CONCEPTS, OLD AND NEW

M. Zimmermann

II. Physiologisches Institut der Universität, Heidelberg, Germany

THE ORIGINS OF TACTILE NEUROPHYSIOLOGY

Modern neurophysiology started when it was possible to record single unit action potentials. Another requirement for good modern neurophysiological work, at least when sensory systems are involved, is that some kind of a quantitative stimulus is used. Both requirements were fulfilled for the first time by work performed 54 years ago by Adrian and Zotterman at Cambridge, published in 1926 as a series of papers in the J. of Physiology. Fig. 1 is a reproduction from their analysis of sensory impulses from the cat's footpad, showing the ingenious apparatus used to produce graded skin deformations (Fig. 1A), the time course of discharge (B) and the intensity function of a "pressure receptor" (C) of the skin. The unit action potentials were amplified for the first time with a vacuum tube amplifier and could therefore be recorded with a capillary electrometer.

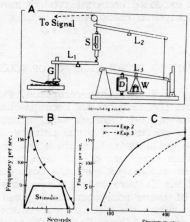

Fig. 1. Investigation of cutaneous mechanoreceptors of the cat, pioneer study of Adrian and Zotterman (1926). A: Stimulator for reproducible and quantitative mechanical skin stimuli. B: Time course of discharge of pressure receptor. C: Transformation of stimulus intensity into discharge frequency of pressure receptors.

Many fundamentally important observations were made by the authors in this now
historical paper. For example, they found that there was no change in the size of
unit action potential when pressure to the skin was varied (Fig. 2), a statement
which was not trivial at that time - as it would seem today! The authors concluded
that rather than the amplitude of the nerve impulse, the discharge frequency of
receptors must be considered the carrier of the information on stimulus intensity.

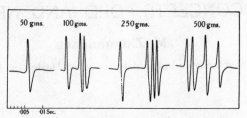

Fig. 5. Analysis of electrometer records, *Exp.* 2, showing that the size of individual
impulses does not vary with the stimulus.

Fig. 2.   Action potentials of a single cutaneous mechano-
          receptor at different intensity of stimulus
          (indicated in gram weight). From Adrian and
          Zotterman (1926).

This notion was further supported when they plotted discharge frequency of pressure receptors versus strength of mechanical skin stimulation (Fig. 1C). There is
a monotonic function which is a prerequisite for the unambiguous decoding of
information on stimulus strength from the impulse discharge. Had the authors plotted their data on a double logarithmic coordinate system, they would have discovered Stevens' law of power functions which was formulated about 40 years later.

These pioneer studies were extended and confirmed by many subsequent investigators.
Never was there a contradiction to these early findings which opened the scene of
modern neurophysiology.

THE PRESENT STATE OF NEUROPHYSIOLOGY OF TOUCH

What is the situation today in skin mechanoreception? First, a clear correlation
exists between structure and function of cutaneous mechanoreceptors. Fig. 3 shows
a survey of mechanoreceptors of the glabrous skin. Histologically there are 4 types
of nerve endings: Meissner's corpuscle (in the primate; the homologue in the cat is
Krause's endbulb), Merkel's cell, Ruffini's organ and the Pacinian corpuscle.
These 4 basic types of mechanoreceptors are the sensors for our tactile system.
The brain creates from their discharges even the most complex haptic perception
when, for example, we move our fingers over a three dimensional object.
The functional properties of these receptors can be best analyzed during a mechanical ramp stimulus which indents the skin at a constant velocity (Fig. 3). The
SA receptors (SA = slowly adapting) have been correlated with the Merkel cell
complex (SA I) and Ruffini type endings (SA II). SA II often exhibits a particular
sensitivity to stretching the skin parallel to the surface. The SA receptors might
be considered as sensors of steady-state skin deformations, their rate of discharge
being monotonically related to the intensity of stimulation (Fig. 4). The RA receptor (RA = rapidly adapting), histologically identified as Meissner's corpuscle

Survey on cutaneous mechanoreceptors

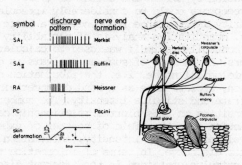

Fig. 3.   Cutaneous mechanoreceptors (schematic); discharge
characteristics (left) and structure of nerve
ending (right). From Zimmermann (1978).

(in primates) or Krause's bulb (in the cat), measures the velocity of the movement
(Fig. 5). The PC receptor (PC = Pacinian Corpuscle) often responds to transients
of movement which contain changes in velocity, hence it has been designated an
acceleration detector. SA I and SA II receptors are today's names for what Adrian
and Zotterman (1926) called pressure receptors, while RA and PC receptors constitute their touch receptors. Examples of discharge characteristics of SA I and RA
receptors are shown in Figs. 4 and 5, respectively.

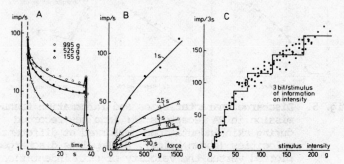

Discharge of the SA-receptor, 42 and 52 years after Adrian's and
Zotterman's pioneer work. A, B from Jänig, Schmidt & Zimmermann 1968
C from Zimmermann 1978

Fig. 4.   Functional characteristics of cutaneous slowly
adapting mechanoreceptors (SA receptors). A: Time
course of discharge frequency of an SA receptor at
different forces (indicated in g, gram weight)
exerted onto the large footpad of the cat. B: Intensity characteristics of same unit as in A,
measured at various times after onset of pressure
stimulus. C: Intensity characteristics of another
SA receptor. Each dot is the result of a pressure
stimulus 3s in duration. The staircase drawn into
the field of scatter of measurements yields a
figure of 3 bit per stimulus of information
intensity.

THE BIOLOGICAL NOISE IN RECEPTOR DISCHARGES

A most important observation by Adrian and Zotterman (1926) was that the rate of
discharge of pressure receptors could be considerably irregular. It was recognized
later that the irregularity in discharge frequency is a basic principle of nervous
function. More recently, quantitative analysis of variability in the receptor
discharge to precisely controlled stimuli was used to calculate the information
transmission in receptors. An example of such an analysis is given in Fig. 4C for
an SA I receptor. The biological noise, i.e. the fluctuation of discharge rate
upon repeated stimulation, appears here as a band of uncertainty in the relationship between discharge rate and stimulus intensity. The number of distinguishable
states of stimulus intensities in this blurred relationship can be estimated by
drawing a staircase function into the field of scatter of experimental points.
This yields about 8 levels of stimulus intensity, information on which can be
deduced from the receptor discharge. In terms of information theory, the measure
of information (I) per stimulus contained in the afferent discharge of this
SA receptor is calculated as

$$I = \log_2 8 = 3 \text{ bits per stimulus.}$$

Apart from this graphic procedure, several mathematical approaches have been used
to calculate the information content of the receptor discharge (e.g. Werner and
Mountcastle, 1965; Kruger and Kenton, 1973).

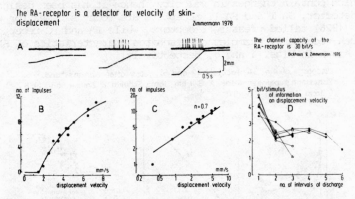

Fig. 5.   Discharge characteristics and information trans-
           mission in RA receptors. A: Impulses recorded
           during skin indentations performed at different
           velocities (shown below). B: Impulses discharged
           plotted versus the velocity of skin indentation.
           C: Same as B, however plotted in double logarith-
           mic coordinate system. D: Information content on
           displacement velocity in 6 different RA receptors
           plotted against number of impulse intervals taken
           into account for the calculation. A, B, C from
           Zimmermann (1978); D from Dickhaus and Zimmermann
           (1976).

A similar analysis was applied to the discharge of RA receptors (Dickhaus, 1976),
the variability in the intervals between successive impulses in a discharge being

used to express the fluctuations. When the first interval in each discharge to the linearly moving stimulus was used for this analysis, between 3.5 and 4.7 bit of information on displacement velocity were obtained (Fig. 5D); that is, in the first interspike interval of a single RA afferent up to 26 levels of indentation velocity can be discerned. However, when a temporally increasing number of subsequent interspike intervals in each discharge was included in the calculation, the information content markedly decreased (Fig. 5D).

The maximum rate of information transmission, or channel capacity (in bits per s), in RA receptors was evaluated by increasing the stimulus repetition rate to as much as 10 Hz in experiments such as those shown in Fig. 5. Values of 30 bits per s were obtained, which are extremely high when compared with the figure of .5 to 10 bits per s estimated for the SA receptors (Werner and Mountcastle, 1965). Thus, the RA receptor is well suited for providing information to the central nervous system on a parameter of moving stimuli. This might be considered an indication that the RA receptor is of high functional significance for the tactile system: it is really the receptor for touch as suggested 53 years ago by Adrian and Zotterman.

PSYCHO-NEURONAL CORRELATIONS

Major progress in research on the tactile system was provided by applying the single unit recording method to human nerves by Hensel and Boman (1960) and Hagbarth and Vallbo (1968). Vallbo and associates have analyzed the discharges of single human RA and SA receptors. They could correlate these discharges directly with the sensation reported by the subjects simultaneously. One of their findings was that a single impulse in a single RA receptor of the fingertip regularly provoked a conscious sensation (Fig. 6A). This was not the case when the palm of the hand was stimulated (Fig. 6B); here, considerable spatial summation was obviously required from many RA receptors to evoke a sensation. This was also true for SA receptors, independent of the location.

Correlation of discharge of a cutaneous RA-fibre (o)
with detection threshold (•) in human subject, in

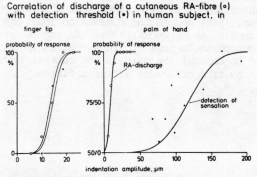

Fig. 6.   Thresholds in human subjects of RA receptors and
of detection of the mechanical stimulus. The or-
dinates plot the probability of occurence of a
nerve impulse in single RA afferents (o) and of
a yes response (•) of a detection task in the
same subject, recorded simultaneously, when the
stimulus intensity (abscissa) of successive stimuli was varied at random. Data were from measurements at fingertip (left graph) and palm of
hand (right graph). From Vallbo and Johansson
(1976).

Another outcome of this work relates to the spatial discriminative capacity of the human hand, which can be quantified by measuring two-point-threshold (Weber, 1835; Weinstein, 1968). Johansson and Vallbo (1976) established that two-point-threshold at various sites on the human hand was inversely related to the density of inner-vation by RA and SA I afferents rather than to the size of the receptive fields of these afferents. Other important correlations between the neurophysiological characteristics of cutaneous mechanoreceptors and psychophysical determinants of the human tactile system came from work by Mountcastle and collaborators, who used sinusoidal mechanical skin deformations at various frequencies in human subjects and monkeys. Two major findings emerged from these studies: (1) Low frequency (around 30 Hz) and high frequency (around 250 Hz) stimulation evoke qualitatively different sensations, termed flutter and vibration, respectively. From comparisons of psychophysical thresholds with the thresholds of receptor discharges, it could be established that flutter sensation is due to excitation of the RA receptors, whereas the sensation of vibration depends on the excitation of PC afferents. (2) Temporal discrimination, measured as the capacity to detect differences in the frequency of sinusoidal stimulation, was much better in the low frequency range (flutter, RA receptors) than in the high frequency range (vibration, PC receptors) It was concluded from this and related studies that the representation of infor-mation in the central nervous system must be different for RA and PC receptors, and only the information from RA receptors enabled the subject to reach the high discrimination capacity for frequency changes.

The PC receptors - acceleration sensors - have the lowest absolute thresholds for fast transients: a vibration of 250 Hz may activate a PC receptor at amplitudes well below 1 µm. Repetitive excitation of a single PC afferent by vibratory stimuli may lead to a conscious sensation in man and monkeys.

PC afferents have a major inhibitory action on the sensory pathways. This could be established in psychophysical experiments using two mechanical skin stimuli. A low amplitude vibration at 300 Hz delivered through one stimulator was particularly effective in elevating the detection threshold for any kind of mechanical skin displacement from the other stimulator (Ferrington, Nail and Rowe, 1977). In neurophysiological investigations in cats it could be shown that the PC afferents produced powerful inhibition of afferent information in the spinal cord (Jänig, Schmidt and Zimmermann, 1968) and in the cuneate nucleus (Bystrzycka, Nail and Rowe, 1977; Jänig, Schoultz and Spencer, 1977). It can be concluded from these psychophysical and neurophysiological findings that an important function of the PC receptors might be to control the gain in the central transmission of sensory information originating from cutaneous afferent fibres.

EFFECTS OF ISCHEMIA IN THE SKIN SENSES

When Yngve Zotterman returned to Stockholm after the golden twenties in England, part of his research was devoted to the question of abnormal sensation during ischemia of the limb. This work was published in his thesis in 1933, the year of my birth. He concluded from his psychophysical experiments, using two plethysmographic cuffs, that nerve impulses produced at the site of compression must be the causes of the tingling and buzzing paresthesiae felt when compression ceased. His suggestion of abnormal impulse generation has been confirmed recently by direct recording from human nerve under the same conditions of experimental ischemia (Fig. 7, Torebjörk, Ochoa and McCann, 1979). Professor Ochoa has presented these findings to this symposium.

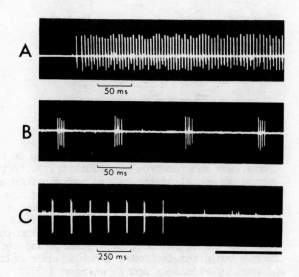

Fig. 7.   Discharges of nerve fibres of human arm after re-
          lease from experimental ischemia. B and C is from
          same unit, which is different from unit in A.
          Various types of paresthesiae were reported by the
          subject during these abnormal discharges. From
          Torebjörk, Ochoa and McCann (1979).

Surface anesthesia of the limb is another feature of ischemia. So far this has been attributed to conduction block in afferent nerves. Animal experiments, however, have shown that the mechanoreceptors of the skin are affected by ischemia long before axonal conduction block occurs (Fig. 8): In both SA and RA receptors of the rat's hindfoot, the excitability of mechanoreceptors by skin deformations dramatically decreased within a few minutes of ischemia. It took much longer for a conduction block to occur in the nerve containing the afferent axons. The long time delay required to block conduction, reflected in the amplitude of the compound action potential, might have been due to the presence of residual circulation in the proximal part of the limb where the compound action potentials were elicited.

A similar finding has been reported by Hensel investigating lingual thermoreceptors. Most other authors concerned with the effects of ischemia have claimed that interruption in the transmission of the neural message was due to axonal block rather than to receptor inexcitability (Frankenhaeuser, 1969).

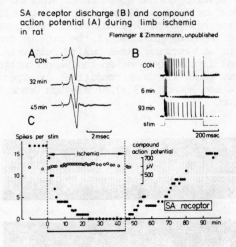

SA receptor discharge (B) and compound
action potential (A) during limb ischemia
in rat
Fleminger & Zimmermann, unpublished

Fig. 8.   Effects by ischemia of the rat's hindleg on nerve
           conduction and receptor excitability. A: Compound
           action potential of the tibial nerve before and
           during ischemia produced by occlusion of the
           iliac artery. B: Discharge of an SA receptor to
           a skin indentation before, during and after
           ischemia. C: Time course of size of action poten-
           tial (o) and of SA receptor discharge (●) before,
           during and after ischemia. Unpublished results
           of work by Fleminger and Zimmermann (1976).

                    NOCICEPTORS FROM 1936 TO 1979

Zotterman's work also strongly influenced the development of current concepts of
pain. His publication "Specific action potentials in the lingual nerve of cat",
which appeared in 1936 in the Skandinavisches Archiv für Physiologie, was a
decisive breakthrough in the neurophysiology of pain. Among the single unit potentials recorded from the cat's lingual nerve, he discerned some slowly conducting
fibres of small amplitude which could be excited only by strong, potentially
damaging stimuli (Fig. 9). These recordings certainly represent the first clear
demonstration of what we would today call a nociceptor. Yngve Zotterman searched
for such nociceptive fibres for many years, and eventually achieved this success
after a long period of failure! He concluded from these results that specific
"pain fibres" exist which convey information on noxious events; this was the first
direct evidence in favour of the specificity theory of pain, a finding which has
been confirmed repeatedly in the past 20 years (Iggo, 1959; Burgess and Perl, 1967;
Beck, Handwerker and Zimmermann, 1974). As an example of contemporary research the
responses to controlled noxious radiant heat of a C fibre nociceptor from the cat's
foot is depicted in Fig. 10.

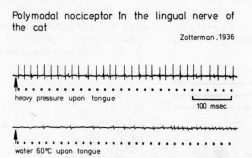

Fig. 9.   Single units recorded from the lingual nerve of
            the cat. The large spike is from a sensitive
            mechanoreceptor. The small spike is from a unit
            responding either to heavy pressure upon tongue
            or to a jet of hot water (60°C) on tongue. This
            unit was therefore called a pain fibre. From
            Zotterman (1936).

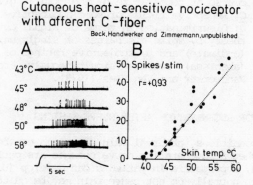

Fig. 10.   Responses of a C fibre nociceptor in the cat's
            footpad to radiant heat. A: Discharges of the
            nociceptor to heat stimuli of different intensi-
            ties, as indicated; the time course of skin tem-
            perature is given below, the duration of the
            heat stimulus was 10 s. B: Number of spikes per
            stimulus plotted versus the temperature of the
            radiant heat stimulus. Linear regression yielded
            a correlation coefficient of r = 0,93.
            Unpublished data from work by Beck, Handwerker
            and Zimmermann (1974).

The most direct proof that nociceptors in fact mediate pain sensation comes from
recent work by Gybels, Handwerker and van Hees (1979). They recorded from human
C fibres during exposure of the skin of the experimental subjects to controlled
graded heat stimuli. The discharge frequency of single C units was highly

28   Manfred Zimmermann

correlated with the subjective rating of heat pain intensity (Fig. 11).

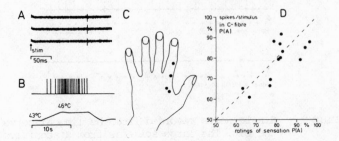

C-fibre discharges in man are correlated with subjective
rating of sensation of noxious heat applied to skin
Gybels,Handwerker & VanHees 1979

Fig. 11. Correlation of C fibre discharges and subjective
ratings of sensation measured simultaneously in
human subject. A: Response to electrical skin
stimulation of C̄ fibre in human superficial
radial nerve. B: Discharge of the unit in A,
transformed to standard pulses, to heating of the
cutaneous receptive field by a thermode; time
course of temperature of the skin is given below.
C: Receptive fields of some nociceptive C fibres.
D̄: Discrimination in terms of signal detection
theory by spike discharges of single units
(ordinate) and by subjective ratings of sensation
(abscissa), recorded simultaneously. From Gybels,
Handwerker and Van Hees (1979).

NOCICEPTOR REGENERATION AFTER NERVE LESIONS

Under pathological conditions, abnormal pain sensation can be elicited from the
skin. For example, when a regenerating nerve reaches its peripheral innervation
territory in the skin, hyperpathia in patients can develop (Head, 1920). Moderately
intense stimuli which normally do not cause pain may be quite painful in such
patients. I recall that in 1975, when Yngve Zotterman was in Heidelberg, he had
just recovered from a broken arm. When testing his skin with radiant heat stimuli
it turned out that the threshold for pain was much lower on the hand of the previously fractured side than on the other hand. His hypothesis was that some nerve
fibres were crushed by the mechanical load of the fracture and subsequent plaster,
and that fibres developed greater sensitivity during recovery. This situation was
simulated in animal experiments done by us at that time (Dickhaus, Zimmermann and
Zotterman, 1976). Cutaneous nerves were crushed in the cat's foot, and subsequently single C fibres were analyzed with regard to their responsiveness to noxious
stimuli. When the sprouting fibres reached the skin, nociceptors with functional
characteristics similar to those in normal skin developed immediately (Fig. 12A).
However, in a sample of regenerated C nociceptors, the average threshold to
noxious skin heating was significantly reduced by 4°C when compared to normal
conditions in control animals (Fig. 12B). This finding might account for hyperpathia to thermal stimuli in patients with regenerated nerves.

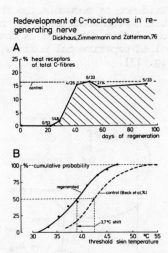

Fig. 12. Regeneration of C fibre nociceptors in the cat.
A: Recovery of the proportion of heat sensitive
C fibres in the cat's plantar nerves after experimental nerve crush. The nerve lesion by crush was
performed about 35 mm from the innervation territory of these nerves. B: The ordinate plots the
number of C fibres excited by radiant heat stimulus to the receptive field at the temperature
given by abscissa. From Dickhaus, Zimmermann and
Zotterman (1976). Control data in A, B are from
Beck, Handwerker and Zimmermann (1974).

Many other conditions of chronic pain might be understood in terms of the altered
behaviour of nociceptors or their afferent fibres   (Zimmermann, 1979).

INHIBITION OF PAIN INFORMATION IN THE CENTRAL NERVOUS SYSTEM

Thus, the early concept promoted by Yngve that pain is a neurophysiological specific system is true as far as the periphery is concerned. However, the experimental
facts necessitate other concepts for the central nervous system. Here we do not
find a specific center for pain: there is neither a neurosurgical operation
available which gives complete and long lasting relief of chronic pain, nor have
neurophysiologists found brain areas which are engaged solely in the processing
of information from nociceptors. Specificity in the central nervous system would
be too simple a concept of nervous system function as a basis for human and
animal behaviour.

A variety of inhibitory actions are known to modulate pain information in the
central nervous system. One of these acts on neurons in the spinal cord which
receive synaptic input from the nociceptive afferent fibres. Yngve Zotterman, in
his thesis (1933), had already considered the possibility that painful messages
might be inhibited in the central nervous system, as could be concluded at that
time from work by Sir Henry Head and Otfried Förster. There is ample proof for this
contention from neurophysiological and behavioural observations during the last
decade of research (for reviews see Liebeskind and Paul, 1977;

Fields and Basbaum, 1978). Evidence is accumulating that multiple descending
systems originating in the brain stem contribute to the modulation of neurons in
the spinal cord. We have studied this situation in the cat. We used radiant heat
to the skin as a well-controlled experimental noxious stimulus, and recorded from
single dorsal horn cells (Fig. 13).

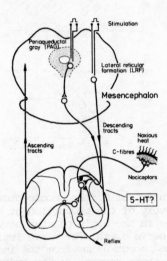

Fig. 13. Descending inhibition originating in the midbrain
of dorsal horn neurons in the spinal cord excited
by noxious skin heating (schematic). Sites of
electrical stimulation are shown in midbrain PAG
and LRF. 5-HT indicates that serotonin probably
is the neurotransmitter substance involved in
this descending inhibition.

The discharge frequencies of the neurons upon graded noxious skin heating were
linearly related to the temperature of the skin stimulus. At a given temperature
of heating, the dorsal horn responses were inhibited by electrical stimulation at
two sites in the midbrain (Fig. 13), the periaqueductal gray (PAG) and the lateral
reticular formation (LRF). However, the two inhibitory systems are differently
organized, as is revealed from the different effects on the characteristics of the
encoding of the intensity of noxious heating (Fig. 14). PAG stimulation produces
a change in slope of the encoding function, which might be interpreted as a change
in gain of the spinal cord transmission system for pain information. On the other
hand, a parallel shift of the encoding function occurs upon stimulation in the LRF,
suggesting a change in setpoint of the transmission system with an increase in
threshold of noxious skin temperature.

These findings are relevant when considering the analgesia obtained by similar
midbrain stimulation in animals and human patients. However, electrical brain
stimulation is a completely non-physiological situation. It would be most important
to know how these descending inhibitory systems can be activated by physiological
or psychological mechanisms, in order to study their role in the control of pain.
This should be a major topic of future research.

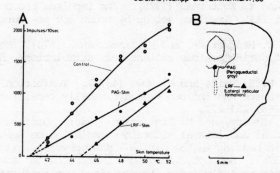

Fig. 14. Two systems of descending inhibition originating
in the midbrain of spinal dorsal horn neurons.
A: Discharges of a dorsal horn unit in anestheti-
zed cat plotted versus the intensity of noxious
skin heating. Measurements were taken during
simultaneous stimulation of the PAG (●) or of the
LRF (▲), or in the absence of electrical midbrain
stimulation (o). B: Location of stimulation electrodes in the midbrain. From Carstens, Klumpp and
Zimmermann (1980).

EPILOGUE

A list of references for work in the somatosensory system should have Yngve
Zotterman in the first and in the last item: Adrian and Zotterman (1926),
Zotterman (1936 to 1976), symbolizing his rich influence on this field of research.
However, this is only a single page in the long book of his scientific life:
there is practically no field in sensory physiology to which he would not have
made contributions of outstanding value. This monograph will hopefully provide
the reader with an impression of Yngve Zotterman's creativity. I wish him a
continuing active and productive life .

ACKNOWLEDGEMENTS

The author expresses his gratitude to Mrs. Almuth Manisali for the graphics,
Ms. Petra Berndt for typing the manuscript and Dr. Earl Carstens for many
suggestions and for improving the English. The author's work referred to in this
article has been supported by the Deutsche Forschungsgemeinschaft .

REFERENCES

Adrian, E. D., and Y. Zotterman (1926). The impulses produced by sensory nerve endings. Part III. Impulses set up by touch and pressure. J. Physiol., 61, 465-483.

Beck. P. W., H. O. Handwerker, and M. Zimmermann (1974). Nervous outflow from the cat's foot during noxious radiant heat stimulation. Brain Res., 67, 373-386.

Bystrzycka, E., B. S. Nail, and M. Rowe (1977). Inhibition of cuneate neurones: its afferent source and influence on dynamically sensitive "tactile" neurones. J. Physiol., 268, 251-270.

Carstens, E., D. Klumpp, and M. Zimmermann (1980). Differential inhibitory effects of medial and lateral midbrain stimulation on spinal neuronal discharges to noxious skin heating in the cat. J. Neurophysiol., (in press).

Dickhaus, H. (1976). Neurophysiologische Untersuchungen sowie system- und informationstheoretische Analyse der Eigenschaften von rasch adaptierenden Mechanorezeptoren der unbehaarten Haut der Katzenfußsohle. Dissertation Technische Hochschule Karlsruhe.

Dickhaus, H., M. Sassen, and M. Zimmermann (1976). Rapidly adapting cutaneous mechanoreceptors (RA): coding variability and information transmission. In Y. Zotterman (Ed.), Sensory functions of the skin in primates, Pergamon Press, Oxford, pp. 45-54.

Dickhaus, H., M. Zimmermann, and Y. Zotterman (1976). The development in regenerating cutaneous nerves of C-fibre receptors responding to noxious heating of the skin. In Y. Zotterman (Ed.), Sensory functions of the skin in primates, Pergamon Press, Oxford, pp. 415-425.

Ferrington, D. G., B. S. Nail, and M. Rowe (1977). Human tactile detection thresholds: modification by inputs from specific tactile receptor classes. J. Physiol., 272, 415-433.

Fleminger, S., and M. Zimmermann (1976). Excitability changes in cutaneous mechanoreceptors of the rat's foot during ischemia. Pflügers Arch.-Eur. J. Physiol., 365, suppl., R 52.

Frankenhaeuser, B. (1949). Ischaemic paralysis of a uniform nerve. Acta Physiol. Scand., 18, 75-98.

Gybels, J., H. O. Handwerker, and J. Van Hees (1979). A comparison between the discharges of human nociceptive nerve fibres and the subject's ratings of his sensations. J. Physiol., 292, 193-206.

Hagbarth, K.-E., and A. B. Vallbo (1968). Activity from skin mechanoreceptors recorded percutaneously in awake human subjects. Exptl. Neurol., 21, 270-289.

Head, H. (1920). Studies in neurology. Oxford University Press, London.

Hensel, H., and K. K. A. Boman (1960). Afferent impulses in cutaneous sensory nerves in human subjects. J. Neurophysiol., 23, 564-578.

Hensel, H. (1954). Das Verhalten der Thermorezeptoren bei Ischämie. Pflügers Arch. ges. Physiol., 257, 371-383.

Iggo, A. (1959). Cutaneous heat and cold receptors with slowly conducting (C) afferent fibres. Q. J. Exp. Physiol., 44, 362-370.

Jänig, W. (1971). Morphology of rapidly and slowly adapting mechanoreceptors in the hairless skin of the cat's hind foot. Brain Res., 28, 217-231.

Jänig, W., R. F. Schmidt, and M. Zimmermann (1968). Single unit responses and the total afferent outflow from the cat's footpad upon mechanical stimulation. Exp. Brain Res., 6, 100-115.

Jänig, W., T Schoultz, and W. A. Spencer (1977). Temporal and spatial parameters of excitation and afferent inhibition in cuneothalamic relay neurons. J. Neurophysiol., 40, 822-835.

Johansson, R. and A. B. Vallbo (1976). Skin mechanoreceptors in the human hand: neural and psychophysical thresholds. In Y. Zotterman (Ed.), Sensory functions of the skin in primates, Pergamon Press, Oxford, pp. 171-184.

Kruger, L., and B. Kenton (1973). Quantitative neural and psychophysical data for cutaneous mechanoreceptor function. Brain Res., 49, 1-24.

LaMotte, R. H., and V. B. Mountcastle (1975). Capacities of humans and monkeys to discriminate between vibratory stimuli of different frequency and amplitude: a correlation between neural events and psychophysical measurements. J. Neurophysiol., 38, 539-559.

Liebeskind, J. C., and L. A. Paul (1977). Psychological and physiological mechanisms of pain. Ann. Rev. Psychol., 28, 41-60.

Mountcastle, V. B., R. H. LaMotte, and G. Carli (1972). Detection thresholds for stimuli in humans and monkeys: comparison with threshold events in receptive afferent fibres innervating the monkey hand. J. Neurophysiol., 35, 122-136.

Torebjörk, H. E., J. L. Ochoa, and F. V. McCann (1979). Paresthesiae: Abnormal impulse generation in sensory nerve fibres in man. Acta Physiol. Scand., 105, 518-520.

Vallbo, A. B., and R. Johansson (1976). Skin mechanoreceptors in the human hand: an inference of some population properties. In Y. Zotterman (Ed.), Sensory functions of the skin in primates, Pergamon Press, Oxford, pp. 185-199.

Weber, E. H. (1835). Über den Tastsinn. Arch. Anat. Physiol. wiss. Med., 152-159.

Weinstein, S. (1968). Intensive and extensive aspects of tactile sensitivity as a function of body part, sex and laterality. In D. R. Kenshalo (Ed.), The Skin Senses, C. C. Thomas, Springfield, Illinois, pp. 195-222.

Werner, G, and V. B. Mountcastle (1965). Neural activity in mechanoreceptive cutaneous afferents: stimulus-response relations, Weber functions and information transmission. J. Neurophysiol., 28, 359-397.

Zimmermann, M. (1978). Mechanoreceptors of the glabrous skin and tactile acuity. In R. Porter (Ed.), Studies in Neurophysiology. Cambridge University Press, Cambridge, pp. 267-289.

Zimmermann, M. (1979). Peripheral and central nervous mechanisms of nociception, pain and pain therapy: facts and hypotheses. In J. J. Bonica and J. C. Liebeskind (Eds.), Advances in pain research and therapy, Vol. 3. Raven Press, New York, pp. 3-23.

Zotterman, Y. (1933). Studies in the peripheral nervous mechanism of pain. Acta Med. Scand., 80, 1-64.

Zotterman, Y. (1936). Specific action potentials in the lingual nerve of cat. Scand. Arch. Physiol., 75, 105-119.

Zotterman, Y. (Ed.) (1976). Sensory functions of the skin in primates. Pergamon Press, Oxford.

MULTIPLE TACTILE PATHS IN THE NERVOUS SYSTEM OF MAMMALS

G. Gordon

University Laboratory of Physiology, Oxford, UK

ABSTRACT

Much confusion has arisen in recent years about the roles of different tracts and nuclei in sensations of touch. In this essay the origin of the 'classical' view is briefly examined, and early ideas considered along with new evidence about ascending and descending systems discovered and studied in higher animals. It is suggested that these systems should no longer be thought of as isolated elements but as complementary subsystems showing various degrees of functional specialization and continuously interacting with each other.

KEYWORDS

Tactile sensitivity; tactile discrimination; tactile quality; Brown-Séquard syndrome; tactile paths; dorsal columns; recovery from lesions; descending control.

INTRODUCTION

In any subject at a particular time there are one or two books or articles which are going to influence the whole direction taken by that subject over the next generation; and for me Yngve Zotterman's paper on Touch, Pain and Tickling, published in the Journal of Physiology just forty years ago (Zotterman, 1939), is of that kind. Full of experimental facts, it generated ideas some of which have come to fruition and others which remain challenging. By carefully analysing records from what he often calls 'few-fibre' preparations of cutaneous nerves he was able to show that different forms of skin stimulus excite different proportions of the various classes of nerve-fibres and therefore presumably of specialized sense-organs. He concluded that these differences underlie the qualities of the sensations we feel even if we very often include them all casually within the general category of touch. Some sorts of touch have an element of tickle: others have not. When I read this paper again with the knowledge that I would be speaking on this very splendid occasion it occurred to me that one fact mentioned in it - that is, the existence in the central nervous system of more than one tactile pathway - might be an appropriate subject to re-examine. So I propose to take a look backwards at the origin of this discovery, and see how new knowledge, derived from human cases and from anatomical, electrophysiological and behavioural work on animals, has changed our ideas about it.

CLINICAL EVIDENCE

The idea is an old one. As far as I can tell it arose out of a study of a large
group of clinical cases of injury or disease of the spinal cord which showed what
used to be called 'Brown-Séquard paralysis'. Their primary characteristic was the
existence of paresis on one side − the side of the lesion − and analgesia,
coupled with loss of temperature sense, on the other. Brown-Séquard had been equi-
vocal in his account of disturbances of touch; but Petrén (1902) and shortly after-
wards Head and Thompson (1906) examined this question critically and came to the
conclusion that there must be two pathways concerned with touch. By using a number
of quantifiable sensory tests Head and Thompson found that in all cases in which
the sense of touch remained, there was a striking dissociation between tactile sensitivity, which remained at its normal threshold everywhere, and spatial discrimination (measured as a two-point threshold by E.H. Weber's compass test) which was
seriously impaired on the paretic side. Ability to recognise the position in which
a limb was placed was often but not always disturbed on that side.

Although post-mortem evidence was not usually available, this clinical evidence,
interpreted in terms of the anatomical knowledge of that time, led to the well-
known conclusion that in unilateral damage to the spinal cord the motor loss was
attributable to damaged cerebrospinal systems that had already crossed, the loss
in spatial discrimination and any proprioceptive disturbance to injury of the only
long ipsilateral ascending tract then known − the dorsal or posterior column −
and preservation of tactile sensitivity on that side to a tactile component in the
spinothalamic tract which, because it had crossed early, was out of harm's way.
The tactile sensitivity sustained by this spared spinothalamic tract has since had
attributed to it the mediation of 'coarse touch' and the dorsal column became the
seat of 'fine touch'. Whatever the source of the expression 'coarse touch' in this
context, it seems a quite inappropriate description of a sense which an observer
as good as Gordon Holmes (1915) described as being intact everywhere and which Head
had shown to be of normal threshold; but it has got into many medical students'
vocabulary. In fact the distinction between the two sides in these cases could
only be detected by two-point testing and possibly by finding some loss of accur-
acy in localizing stimuli.

In man, obviously the best subject for these studies, lesions do not respect anatomical boundaries. Histological evidence is often not available; and in progressive disease post-mortem evidence would in any case be worth little because the
condition at death would be different from that during sensory testing. In non-
progressive disease − and this applies also after experimental lesions in animals − there is a strong tendency towards functional recovery, in which remaining systems are perhaps playing their part by learning new tricks. It is therefore very difficult to determine the roles of particular tracts and nuclei in con-
tributing to the sense and quality of touch. In parenthesis the advantage of
seeing the early as well as the later effects of a lesion is very strikingly illustrated by some of the cases of Brown-Séquard paralysis examined by Gordon Holmes
(1915, 1919) within hours of their injury in the First World War, and whose dominating symptom during the first few days, sometimes for a week or two, was violent
pain referred down the paretic side, a pain easily set off by touch and movement,
and horribly intense. Recovery was complete. We ought not to be surprised at
this symptom now − in fact we should perhaps expect some degree of release from
pain-inhibiting control when descending fibres of the dorsolateral white matter
have been damaged. The gradual recovery might also be expected because these
descending systems, certainly in the monkey, are bilaterally distributed (see

Willis, Haber and Martin, 1977; for further bibliography see Willis and Coggeshall, 1978).

Head and Thompson's attributing the defects in two-point threshold to the dorsal column was perfectly reasonable at that time. No evidence has in my view been clear-cut enough to contradict it, even up to the present — which is not to say they were wholly right. The cases of Cook and Browder (1965) have often been quoted as denying any important sensory role to the dorsal columns; they made surgical lesions of either the cuneate or the gracile funicle in patients with phantom limb pain and found only minimal sensory disturbances, mostly very short-lived. But to quote them in this way results from a misreading or misinterpretation of their paper. Only one of their eight patients had any part of the relevant distal limb preserved, a woman who retained her thumb and two fingers; and she had a permanent deficit in two-point discrimination there. It is in the extremities that this discrimination is highest, and also, as Holmes' (1915) cases show very strikingly, where it is most vulnerable to disturbance. There was no histological verification of these lesions because all the patients survived; and this leaves a small margin of uncertainty even in this case which supports the classical view because some damage may conceivably have been done to the neighbouring dorsolateral fascicle. These limits in clinical investigation make it difficult to refute the challenging proposal by Wall (1970) that the dorsal column is not itself a sensory tract, but rather is concerned in steering other systems to acquire sensory information — not, anyway, if we consider only human evidence. The fact that direct repetitive electrical stimulation of the columns produces a sensation of tingling (see e.g. Shealy, Mortimer and Hagfors, 1970) is not relevant here because this stimulation will act antidromically as well as orthodromically on the dorsal column fibres, so that other ascending systems will be excited through the segmental collaterals of these fibres. One important fact that we do know is that a human patient could detect and localize touch bilaterally after the whole cord except one ventrolateral corner had been demonstrably divided by a knife wound (Wall and Noordenbos, 1977), which is compatible with this region carrying the other tactile pathway of the classical theory.

EXPERIMENTAL EVIDENCE FROM ANIMALS

It is in the structure and properties of ascending tactile systems in animals, and the deficits in learned behaviour produced by selected lesions, that we really see a harvest of new facts; we also face inevitable difficulties in making use of them to understand the mechanisms underlying human sensation. I shall be considering the advances over a span of about fifteen years.

Examined electrophysiologically the dorsal columns and dorsal column nuclei seem splendidly adapted to their allotted role in spatial discrimination: the columns bring a vast input of mainly rapidly-adapting tactile fibres into the nuclei; the cells projecting from the nuclei to the thalamus have small sensitive receptive fields in skin and subcutaneous tissue; and these fields are nearly all constructed on a centre-surround basis which like that of retinal ganglion cells is held to be a basis for increasing spatial and temporal constrast, in this case with excitatory centres and inhibitory surrounds (see e.g. Gordon and Jukes, 1964). So it came as a surprise when it was shown, in cats, dogs and monkeys, that section of the dorsal column had only marginal and temporary effects on performance of a number of discriminatory tasks or on capacity to learn them. Defects certainly appeared, but behavioural performance came back to normal within a month or two. The more demanding the discriminatory task the longer this recovery took; but it happened just the same. One interesting piece of work by Vierk (1973) suggested the sort of adaptation that may have been going on during this period and at the same time offers a method for interpreting this sort of experiment. Having taught

monkeys to discriminate between passively applied discs of various sizes, he sectioned the dorsal column of one side, which produced the usual severe deterioration
on that side, followed by several months of recovery. When the column of the other side was then cut a similar defect appeared but recovery was significantly
faster, suggesting that the monkey had learned to use new cues to discriminate
between the discs and that on the second occasion adopting the new strategy was
that much easier and quicker. This kind of reasoning suggests a partial and very
specialized role for the dorsal columns and their nuclei in discriminatory tasks,
not a total and exclusive one; and this is how many of us now see them. One would
expect that specialized tasks could be found for which their presence was actually
necessary; and Vierk (1974) has identified one of these, namely the ability to
detect the direction of passive movement of an object across the skin, which is
never recovered after section of the columns (see also Brinkman and Porter, 1978).
Their involvement in more complex spatiotemporal tasks is emphasized by the complete loss of ability to discriminate cut-out shapes after dorsal column section
(Azulay and Schwartz, 1975) – an active recognition test containing some components of stereognosis. Other aspects of stereognostic recognition may not be
affected (Vierk, 1978).

This is a field of work where asking the right experimental question is vital,
and it seems in retrospect that most of the earlier experiments had simply failed
to do that, probably in the expectation that cutting so large a tract would be
certain to produce devastating deficiencies. A more rational attitude might be
that it is the most difficult analytical tasks that need the largest number of
fibres and that sensory problems of less complexity can be handled by smaller and
less sophisticated afferent systems.

There are some aspects of motor behaviour which are sensitive to dorsal column
lesions, which is not surprising because the columns supply the motor cortex with
its fast tactile and kinaesthetic feedback (from the monkey's forelimb, Brinkman
and Porter, 1978; from the human hand, Marsden and others, 1977), and also a considerable part of the cutaneous input to the postcentral gyrus (Dreyer and others,
1974) which is presumably used in the sequential generation of exploratory movements. But here again the contribution is partial and specialized; and it is not
easy, as Vierk (1978) puts it in his illuminating analysis of this subject 'to
unveil the relevant component as distinct from a generalized inability to handle
complexity'. One specific defect he quotes in monkeys which are otherwise using
their hands well is in opposing the sides of their thumb and forefinger rather
than the tips as they normally do.

But what other path or paths are involved in the less 'complex' sensory components
of tactile discrimination? A finger was pointed at the dorsolateral fascicle of
the spinal white matter when Morin described the spinocervical tract, an ipsilateral second-order tract which ascends there and relays in the lateral cervical
nucleus, which in turn projects to thalamus and cortex (Morin, 1955; Morin and
Catalano, 1955). Morin's path is mainly made up of fibres with small tactile
receptive fields; it has not got the centre-surround mechanisms of the dorsal
column nuclei; and multimodal convergence on to its cells of origin gives it a
wider dynamic range in its response to peripheral stimuli. This dynamic range
can be drastically reduced by cerebrofugal inhibition, converting this to a purely
tactile tract (Brown, 1971). This part of the cord also carries second-order
fibres to the dorsal column nuclei (Dart and Gordon, 1970, 1973; Gordon and Grant,
1972), and these were also found to be predominantly tactile. It is quite possible
that these fibres, which originate in substantial numbers from levels at least as
low as L 5, caudal to any cells of Clark's column (G. Gordon and G. Grant, unpublished observations), are in fact branches of spinocervical fibres. It is worth
noting that the second-order fibres which ascend in the dorsal columns have similar functional properties to spinocervical fibres (Uddenberg, 1968; Angaut-Petit,

1975) and that their cells of origin may be functionally related to spinocervical cells (Jankowska, Rastad and Zarzecki, 1979). In fact the cells of origin of all these dorsal second-order tracts lie in the same part of the dorsal horn, mainly in Rexed's lamina IV (Craig, 1976; Rustioni and Kaufman, 1977), which is a circumstantial point in favour of close functional relationship. In the monkey second-order cells projecting to the dorsal column nuclei have a similar origin but are more numerous than in the cat (Rustioni, Hayes and O'Neill, 1979). Unfortunately nothing is known either structurally or functionally about these cells in man. Furthermore the only criterion for the presence of a spinocervical tract in man depends on the existence of a recognizable lateral cervical nucleus. The nucleus has been observed, with spinocervical fibres degenerating in it after injury; but apparently it is not constantly present (see Truex and others, 1965; Kircher and Ha, 1968) so it remains an open question how important this nucleus can be for human sensation. One component of the monkey's spinothalamic tract has receptive properties similar to those of the spinocervical tract (Willis and others, 1974), and this is undoubtedly a prominent tract in man.

In spite of the difficulties I have mentioned in establishing some specialized roles for the dorsal column in animal experiments, the importance of the ascending paths in the dorsolateral fascicle has never been in doubt since the question was first investigated. Norrsell (1966, 1975) showed early on that a tactile conditioned reflex where the stimulus was applied to one or other hind leg of a dog, and the response expressed by the dog pushing a button with its nose, was transiently impaired on one side by small lesions of the dorsolateral fascicle of that side; it was not affected by section of one or both dorsal columns; but it was severely and usually permanently affected by cutting both. An interesting shared relationship between these two dorsal pathways has also been shown in experiments on roughness discrimination in cats (Kitai and Weinberg, 1968) and in Vierk's studies on tactile discriminatory capacity in monkeys (Vierk, 1973, 1974): particular defects from which recovery is possible after section of one tend to become permanent after section of both. Taken together with the various detailed examples I have given, this suggests that these two regions operate in a complementary way in analysing the different facets of complex tactile discriminations.

Can we specify, for the experimental animal at least, what actual tracts or connections are involved? In spite of knowing in some detail about the main ascending paths in these two areas, I think the answer is still that we can not. I say that because of the probability that has now come to light that the two are reciprocally interconnected. I have already mentioned ascending connections from the dorsal horn to the dorsal column nuclei of which some make use of the dorsolateral fascicle. We now know of a quite massive system of cells in the gracile and cuneate nuclei which project downwards, some down the dorsolateral fascicle and some down the dorsal column, and end in the dorsal horn. This rather startling fact emerged in some experiments in which Dart and I (Dart and Gordon, 1970, 1975) were recording from cells in the cat's dorsal column nuclei after the dorsal columns had been dissected away at a high cervical level. Apart from the ascending input to the nuclei through the dorsolateral fascicle we found responses that Dart (1971) was able to show by the impulse collision technique to be antidromic, in response to stimuli applied to the fascicle lower down the cord. Since then horseradish peroxidase has been injected into various parts of the spinal cord and this has resulted in labelling by retrograde axonal transport of cell-bodies in and on the borders of these nuclei (Kuypers and Maisky, 1975, 1977; Burton and Loewy, 1977). We have recently enlarged on specific aspects of this work (Armstrong and others, 1979) and have found not only the large ipsilateral projection down the dorsal columns described by these authors but also a substantial component going down the dorsolateral fascicle: we have also shown, for the cat, that no other pathways are involved. Some of these cells are shown in Fig. 1. In the cat, individual descending cells have been shown to receive excitatory inputs from the dorsal

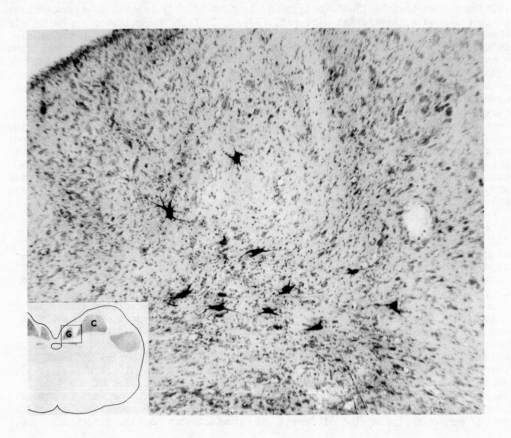

Fig. 1. Transverse section of medulla of cat 1-2 mm caudal to obex, 50 μm
thick. Field centred on right gracile nucleus, showing cells labelled
with horseradish peroxidase 3 days after injection into dorsal columns at
T 13. Hanker-Yates substrate. Lightly counterstained with gallocyanin.
Inset shows position of field within total transverse section. G, gracile
nucleus; C, cuneate nucleus. Width of field, 1.3 mm.

column, from the dorsolateral fascicle, or from both: a few were also excited by
stimulating the contralateral sensorimotor motor cortex (Kleider, 1974). The
unexplored question of their peripheral receptive fields will be important as a
step in establishing their function. The regions of termination of their axons
are not fully known. Autoradiographic studies involving orthograde transport of
tritiated amino-acids injected into the dorsal column nuclei in the cat suggest
that at the level of the brachial enlargement the main termination is in lamina
V of the dorsal horn (Burton and Loewy, 1977), a lamina containing both spinocervical cells (Craig, 1976) and spinothalamic cells (Trevino and Carstens, 1975)
at this level. There is therefore good reason to associate this descending projection with the control of ascending activity in other sensory tracts; but any
more precise suggestion is quite unjustified at this stage.

COMMENT

The evidence from animal experiments may suggest that spinomedullary tactile mechanisms are much more complicated than was envisaged in the early dual concept of
tactile paths evolved from clinical evidence. To some extent this is inevitable
because early ideas could not take into account the existence of descending pathways which could modify ascending activity; so a lesion, certainly anywhere in
the dorsal part of the cord, will have less easily specifiable effects than the
dual concept could allow for. On the other hand the evidence from animals is now
sufficiently extensive and varied to let us shed finally the obstructive assumption that we must have a nominated function for each tract or a tract for each
function, unjustified simplifications which have emerged since the dual theory,
which itself has a lot of essential truth, was first put forward. It is now clear
that these tactile mechanisms are represented by two or more systems in equilibrium, and that the disturbances in the various discriminatory aspects of touch
after lesions must be the result, not of cutting unique channels like telephone
wires, but of a complex process of disequilibration.

REFERENCES

Angaut-Petit, D. (1975). The dorsal column system: II. Functional properties and
bulbar relay of the postsynaptic fibres of the cat's fasciculus gracilis.
Exp. Brain Res. 22, 471-493.
Armstrong, R, Blesovsky, L., Corsiglia, R. and Gordon, G. (1979). Descending
projections from the cat's dorsal column nuclei. J. Physiol. (Lond.). Proceedings; in press.
Azulay, A., and Schwartz, A.S. (1975). The role of the dorsal funiculus of the
primate in tactile discrimination. Exp. Neurol. 46, 315-332.
Brinkman, J., and Porter, R. (1978). Movement performance and afferent project-
ions to the sensorimotor cortex in monkeys with dorsal column lesions. In
G. Gordon (Ed.), Active Touch. Pergamon Press. pp. 119-137.
Brown, A.G. (1971). Effects of descending impulses on transmission through the
spinocervical tract. J. Physiol. (Lond.) 219, 103-125.
Burton, H., and Loewy, A.D. (1977). Projections to the spinal cord from medull-
ary somatosensory relay nuclei. J. comp. Neurol. 173, 773-792.
Cook, A.W., and Browder, E.J. (1965). Functions of posterior columns in man.
Arch. Neurol. (Chicago) 12, 72-79.
Craig, A.D. (1976). Spinocervical tract cells in cat and dog, labelled by the
retrograde transport of horseradish peroxidase. Neurosci. Lett. 3, 173-177.
Dart, A.M. (1971). Cells of the dorsal column nuclei projecting down into the
spinal cord. J. Physiol. (Lond.) 219, 29-30P.
Dart, A.M., and Gordon, G. (1970). Excitatory and inhibitory afferent inputs to
the dorsal column nuclei not involving the dorsal columns. J. Physiol. (Lond.)
211, 36-37P.
Dart, A.M., and Gordon, G. (1973). Some properties of spinal connections of the
cat's dorsal column nuclei which do not involve the dorsal columns. Brain
Res. 58, 61-68.
Dart, A.M., and Gordon, G. (1975). Some properties of spinal connections of the
dorsal column nuclei that do not involve the dorsal columns. In H. Kornhuber
(Ed.), The Somatosensory System. Thieme: Stuttgart. pp. 176-181.
Dreyer, D.A., Schneider, R.J., Metz, C.B., and Whitsel, B.L. (1974). Differential
contributions of spinal pathways to body representation in postcentral gyrus
of Macaca mulatta. J. Neurophysiol. 37, 119-145.
Gordon, G., and Jukes, M.G.M. (1964). Dual organization of the exteroceptive
components of the cat's gracile nucleus. J. Physiol. (Lond.) 173, 263-290.
Gordon, G., and Grant, G. (1972). Afferents to the dorsal column nuclei from the
dorsolateral funiculus of the spinal cord. Acta physiol. scand. 84, 30-31A.

Head, H., and Thompson, T. (1906). The grouping of afferent impulses within the
    spinal cord. Brain 29, 537-741.

Holmes, G. (1915). Spinal injuries of warfare. III. The sensory disturbances in
    spinal injuries. Brit. med. J. ii, 855-861.

Holmes, G. (1919). Pain of central origin. Contributions to Medical and Biolog-
    ical Research, dedicated to Sir William Osler. Vol. 1, pp. 235-246. Hober.

Jankowska, E., Rastad, J., and Zarzecki, P. (1979). Segmental and supraspinal
    input to cells or origin of non-primary fibres in the feline dorsal columns.
    J. Physiol. (Lond.) 290, 185-200.

Kircher, C., and Ha, H. (1968). The nucleus cervicalis lateralis in primates,
    including the human. Anat. Rec. 160, 376.

Kitai, S.T., and Weinberg, J. (1968). Tactile discrimination study of the dorsal
    column-medial lemniscus system and spinocervico-thalamic tract in cat. Exp.
    Brain Res. 6, 234-246.

Kleider, A. (1974). A functional study of some inputs and outputs of the cat's
    dorsal column nuclei. D.Phil. thesis: Oxford.

Kuypers, H.G.J.M., and Maisky, V.A. (1975). Retrograde axonal transport of horse-
    radish peroxidase from spinal cord to brain stem cell groups in the cat.
    Neurosci. Lett. 1, 9-14.

Kuypers, H.G.J.M., and Maisky, V.A. (1977). Funicular trajectories of descending
    brain stem pathways in cat. Brain Res. 136, 159-165.

Marsden, C.D., Merton, P.A., Morton, H.B., and Adam, J. (1977). The effect of
    posterior column lesions on servo responses from the human long thumb flexor.
    Brain 100, 185-200.

Morin, F. (1955). A new spinal pathway for cutaneous impulses. Amer. J. Physiol.
    183, 245-252.

Morin, F., and Catalano, J.V. (1955). Central connections of a cervical nucleus
    (nucleus cervicalis lateralis of the cat). J. comp. Neurol. 103, 17-32.

Norrsell, U. (1966). The spinal afferent pathways of conditioned reflexes to cut-
    aneous stimuli in the dog. Exp. Brain Res. 2, 269-282.

Norrsell, U. (1975). Central nervous structures participating in the simple rec-
    ognition of touch. Acta Neurobiol. Exp. 35, 707-714.

Petrén, K. (1902). Ein Beitrag zur Frage vom Verlaufe der Bahnen der Hautsinne im
    Rückenmarke. Skand. Arch. Physiol. 13, 9-98.

Rustioni, A., and Kaufman, A.B. (1977). Identification of cells of origin of non-
    primary afferents to the dorsal column nuclei of the cat. Exp. Brain Res. 27,
    1-14.

Rustioni, A., Hayes, N.L., and O'Neill, S. (1979). Dorsal column nuclei and asc-
    ending spinal afferents in macaques. Brain 102, 95-125.

Shealy, C.N., Mortimer, J.T., and Hagfors, N.R. (1970). Dorsal column electro-
    analgesia. J. Neurosurg. 32, 560-564.

Trevino, D.L., and Carstens, E. (1975). Confirmation of the location of spino-
    thalamic neurons in the cat and monkey by the retrograde transport of horse-
    radish peroxidase. Brain Res. 98, 177-182.

Truex, R.C., Taylor, M.J., Smythe, M.Q., and Gildenberg, P.L. (1965). The lateral
    cervical nucleus of cat, dog and man. J. comp. Neurol. 139, 93-104.

Uddenberg, N. (1968). Functional organization of long, second-order afferents in
    the dorsal funiculus. Exp. Brain Res. 4, 377-382.

Vierk, C.J., Jr. (1973). Alterations of spatio-tactile discrimination after les-
    ions of primate spinal cord. Brain Res. 58, 69-79.

Vierk, C.J., Jr. (1974). Tactile movement detection and discrimination following
    dorsal column lesions in monkeys. Exp. Brain Res. 20, 331-346.

Vierk, C.J., Jr. (1978). Interpretations of the sensory and motor consequences of
    dorsal column lesions. In G. Gordon (Ed.), Active Touch. Pergamon Press.
    pp. 139-159.

Wall, P.D. (1970). The sensory and motor role of impulses travelling in the dor-
    sal columns towards cerebral cortex. Brain, 93, 505-524.

Wall, P.D., and Noordenbos, W. (1977). Sensory functions which remain in man after complete transection of dorsal columns. Brain 100, 641-653.

Willis, W.D., Trevino, D.L., Coulter, J.D., and Maunz, R.A. (1974). Responses of primate spinothalamic tract neurons to natural stimulation of hindlimb. J. Neurophysiol. 37, 358-372.

Willis, W.D., Haber, L.H., and Martin, R.F. (1977). Inhibition of spinothalamic tract cells and interneurons by brain stem stimulation in the monkey. J. Neurophysiol. 40, 968-981.

Willis, W.D., and Coggeshall, R.E. (1978). Sensory Mechanisms of the Spinal Cord. John Wiley: New York.

Zotterman, Y. (1939). Touch, pain and tickling: an electrophysiological investigation on cutaneous sensory nerves. J. Physiol. (Lond.) 95, 1-28.

CEREBRAL CONTROL OF JAW REFLEXES

S. Landgren and K. Å. Olsson

Department of Physiology, University of Umeå, S-901 87, Umeå, Sweden

ABSTRACT

The effects of electrical stimulation of the cerebral cortex on the monosynaptic jaw closing and the disynaptic jaw opening reflexes were studied in cats anaesthetized with chloralose. The time course of the reflex effects was recorded. Similar rythmic sequences of facilitation and inhibition were observed in both reflexes. The sequence could start with facilitation or inhibition. The latency of the initial effects was short (2.5 ms) indicating a minimum of two synapses in the descending path. The period of the rythmic sequence was approximately 10 ms.

The cortical origin of the effects was located and related to the somatosensory projections, and to the cytoarchitecture. The effects of largest amplitude and most complex time course were evoked from the oral and perioral projections to areas 3a and 3b. Effects evoked from areas 4γ, 5a and 6aβ were less complex and of lower amplitude.

The origin of the cortical effects in cytoarchitectonic area 3a, the function of this area, and the course of the descending paths subserving the short latency effects on the jaw reflexes are discussed. These mechanisms may influence, but presumably do not generate, the normal chewing rythm. It is more likely that they participate in long loop reflexes and feed back control of the jaws. The cortical control of jaw movements is presumably exercized, when the jaws are used in complex behavioral patterns like that of capturing the prey, and not in the chewing or lapping automatisms.

KEYWORDS

Cat; chloralose; jaw reflexes; cerebral cortex; somatosensory projections; cytoarchitecture; cortical area 3a.

INTRODUCTION

Our knowledge of the central nervous mechanisms, which control the jaw movements consists at present of a basic reflexology restricted to the fundamental jaw closing and jaw opening reflexes, a developing analysis of bulbar motor centres, and of observations of jaw movements and other motor responses evoked by stimulation of the

45

hypothalamus or the cerebral cortex. There is a wide scope of development along all of these lines. We have choosen to study the descending control of the jaw reflexes evoked from the cerebral cortex. Information was gained concerning the cortical effects on standardized test reflexes and concerning the organization of the cerebral cortex from which the reflex effects were evoked.

Cortical effects on the jaw reflexes

Let us consider some patterns of behaviour in the cat, which involve jaw movements, and which may be affected by the cerebral cortex.

Chewing is such a pattern which is highly automatic. In the cat chewing mainly consists of a rythmic sequence of opening and closing of the jaws. Due to the type of occlusion of the molars the lower jaw cuts like the branches of a pair of scissors and the temporo-mandibular joint displays a simple hinge function. Lateral movements as well as protrusion and retrusion are restricted. The normal chewing rythm in the cat is 1 - 4 Hz (Morimoto and Kawamura, 1973) and electrical stimulation near the rostral end of the orbital sulcus of the cat's cerebral cortex may evoke chewing with a rythm of 1 - 2 Hz (Morimoto and Kawamura, 1973; Nakamura and co-workers, 1976). Chewing is presumably controlled by lower motor centers in the medulla and the mesencephalon in analogy with the central mechanisms of the stepping movements (Lundberg, 1969; Shick and Orlovsky, 1976).

Lapping is another highly automatic pattern which has been evoked by cortical stimulation as described by Magoun and co-workers (1933) and by Hess and co-workers (1952). Lapping requires coordination of the movements of the jaws, the tongue and the pharynx. Complex bulbar mechanism presumably play an essential role in this coordination. The cerebral cortex may certainly influence chewing and lapping, but it is unlikely that the cerebral cortex exerts the immediate control of the rythmic movements.

An example of jaw movements involved in a less automatic pattern of behaviour are given by the experiments of Dubrovsky and co-workers (1974). The cat sits looking at a rotating wheel, which throws pieces of liver loosely attached to pins on its circumference. The animal jumps to catch the food in the flight. Such jaw movements should require coordination with visual tracking and with movements of extremities and body in the jump. This is a task that may be of a complexity requiring cortical control, and Dubrovsky and co-workers (1974) did in fact observe, that the behaviour was affected by lesions in the anterior sigmoid gyrus.

At present only fragments of information are available concerning the control mechanisms of the above described behavioral patterns. The role of the cerebral cortex in these mechanisms is largely unknown. In all of the three patterns a simple jaw opening and closing movement is integrated. It is therefore fundamental to study the descending paths by means of which the cerebral cortex may influence the final common path to the jaw opening and jaw closing muscles.

According to present views the cerebral cortex facilitates the jaw opening and inhibits the jaw closing muscles. The jaw openers have therefore been classed as flexor and the jaw closers as extensor muscles (Sauerland and co-workers, 1967; Chase and co-workers, 1973; Kubota, 1976). Our results are not in agreement with this generalization. We have studied the monosynaptic jaw closing and the disynaptic jaw opening reflexes and recorded the effects on them evoked by electrical stimulation of the cerebral cortex in the cat anaesthetized with chloralose. The experimental arrangements are shown in Fig. 1.

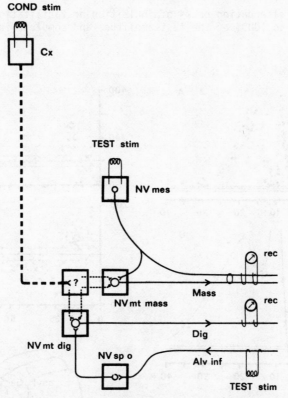

Fig. 1. Diagram showing the experimental set up. The monosynaptic jaw closing reflex evoked by electrical stimulation in the
mesencephalic trigeminal tract (NVmes) and recorded from the
masseteric nerve (Mass). The disynaptic jaw closing reflex
evoked by stimulation of inferior alveolar nerve (Alv inf) and
recorded from the digastric nerve (Dig). Monopolar conditioning
stimulation applied to ipsilateral cerebral cortex (Cx). NVmt=
N motorius n. trigemini; NVspo=N tracturs spinalis trigemini
oralis.

The effects on the reflexes varied with the location of the stimulus on the cerebral cortex, with the stimulus parameters used, and with the level and type of ana-
esthesia (Olsson and Landgren, 1980). The most effective part of the cerbral cortex
was the projection maxima of the maxillary whiskers, the nose and the teeth in areas
3a and 3b of the coronal gyrus. A complex sequence of facilitation and inhibition of
both jaw reflexes was evoked by electrical stimulation of the cortical surface of
these areas using a single anodal pulse or a short train of 3 - 5 pulses (duration:
0.5 or 1 ms, 400 - 600 Hz, threshold strength: 0.3 mA).

The diagrams A - D of Fig. 2 illustrate the effects of such cortical conditioning
on the monosynaptic jaw closing reflex. The sequence of events could start with
facilitation as in Fig. 2 A or with inhibition as in B and C. The latency between
the first cortical shock and the first sign of effect on the masseteric motoneurones,
measured with reference to the arrival of the test volley in the trigeminal motor
nucleus, was short. Minimum latencies of 2.5 ms indicate a disynaptic cortico-trigeminal path as the shortest alternative. The initial facilitation or inhibition

48   Sven Landgren and K.Å. Olsson

was followed by alternating peaks of inhibition or facilitation of large amplitude.
Facilitation up to 1000% of the test amplitude and complete inhibition was often
observed.

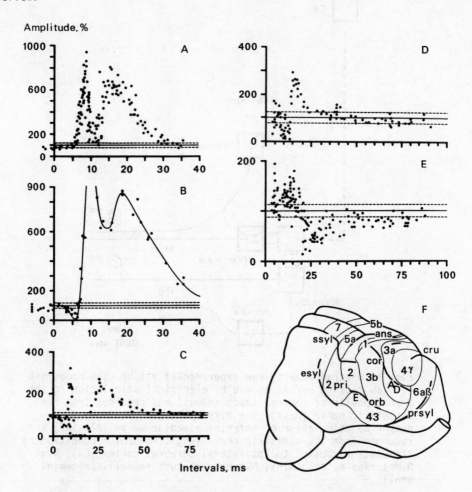

Fig. 2. Effects of cortical conditioning on the monosynaptic
jaw closing reflex. In the diagrams (A-E) the mean and SD of
the unconditioned test amplitude are indicated by the line
drawn in full at 100% and by the broken lines. F: Diagram of
the right hemisphere with indication of the position of the
conditioning stimulus (A, D, E). A standard pattern of the
cytoarchitectonic areas and the nomenclature of the sulci are
given. Plotted curves A-C obtained by the stimulation of point
A in area 3a. Conditioning stimulus: Surface anode; A: 3 pulses, 1 ms, 615 Hz, 1 mA. B: 3 pulses, 1 ms, 615 Hz, 0.8 mA.
C: 1 pulse, 0.5 ms, 2.0 mA. D: 5 pulses, 0.5 ms, 500 Hz,
0.6 mA. E: 3 pulses, 0.5 ms, 444 Hz, 0.8 mA.

The effects on the disynaptic jaw opening reflex were similar to those on the jaw
closing reflex (Olsson and Landgren, 1980). Figure 3 summarizes the observed time
courses of the rythmic events.

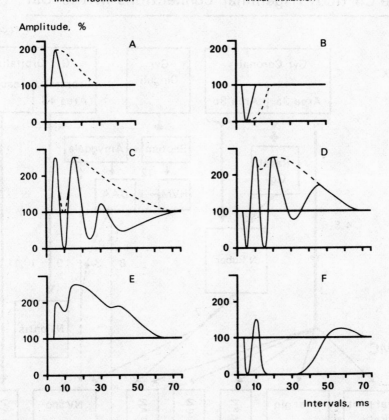

Fig. 3. Schematic illustration summarizing the effects on the
jaw reflexes evoked by electrical stimulation of the cortex.
A-B: Sequences of events starting with short latency facilitation (A) or inhibition (B). C-D: Fully developed sequences
of alternating facilitation and inhibition starting with facilitation (C) or inhibition (D). E-F: Sequences with dominating
facilitation (E) and inhibition (F). Broken lines give alternative time courses.

Initial facilitation or initial inhibition were observed in both types of test reflexes (Fig. 3 A and B). The rythmic pattern could be mainly facilitatory as in
Fig. 3 E or mainly inhibitory as in Fig. 3 F. Rythmic variations displaying sinusoidal curves of the type shown in Fig. 3 C and D were often found. The first 3 - 4
peaks showed a period of approximately 10 ms, but the period was gradually prolonged
and the amplitude decreased with increasing intervals between conditioning and test
shocks.

We interpret our findings as evidence of fast excitatory and inhibitory paths from
the cerebral cortex to the masseteric as well as to the digastric motoneurones. The
two compeeting paths seem to be operated by a switching mechanism which is capable
of choosing between the excitatory and the inhibitory alternative. The location and
functional properties of this switch is unknown.

The Cortico-Trigeminal connections in the Cat

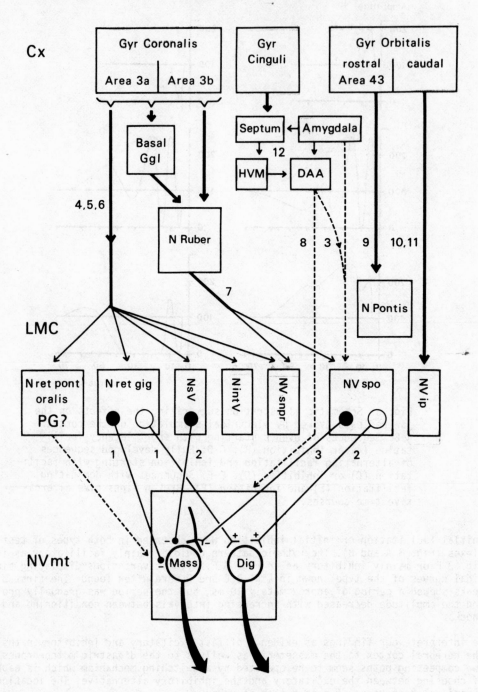

Fig. 4.

The cortico-trigeminal paths

An attempt has been made to summarize the present knowledge concerning the cortico-
trigeminal connections in the diagram of Fig. 4. The evidence of the existence of
the paths shown in Fig. 4 rest on anatomical or electrophysiological experiments
according to the references indicated by figures 1 - 12 in the text of the diagram.

The efferents from the sensory-motor cortex reach the trigeminal motoneurones via
the pyramidal and the cortico-rubro-bulbar tracts. As shown by Wahlberg (1957) cortical lesions do not cause degeneration in the trigeminal motor nucleus itself, but
large presumably fast conducting axons were shown by Wold and Brodal (1973) to approach the trigeminal nuclei from the ventro-medial side and to terminate in the
medial and central parts of the main sensory trigeminal nucleus, thus adjacent to
the motor nucleus.

In the cat the cortical efferents thus do not make monosynaptic connections with the
trigeminal motoneurones. As shown in the diagram of Fig. 4 they terminate on the interneurones of lower motor centres (LMC) in the medulla oblongata. The connections
between these lower motor centres and the final common path are only partially known.
Detailed information is available concerning the connections between the gigantocellular reticular nucleus and the trigeminal motoneurones thanks to the investigations of Nakamura and co-workers (1975 and 1976). They showed that masseteric motoneurones were monosynaptically inhibited and digastric motoneurones monosynaptically
excited by interneurones located in the gigantocellular reticular nucleus. These interneurones were discharged with a shortest latency of 2 ms by electrical stimulation of the coronal or orbital gyri of the cerebral cortex.

The interneurones in the supratrigeminal nucleus which inhibit the masseteric motoneurones were demonstrated by Kidokoro and co-workers (1968b) and the inhibitory and
excitatory interneurones of the oral subdivision of the nucleus of the spinal trigeminal tract were suggested by Bonvallet and Gary-Bobo (1975) and Kidokoro and co-
workers (1968a).

In addition to the pyramidal and rubral paths the trigeminal motoneurones may be
influenced from the hypothalamus as shown by Landgren and Olsson (1977). The course
of this descending path is not investigated in detail, but it functions also after
ablation of the cerebral cortex. Descending paths from the amygdala are indicated

Fig. 4. Diagram showing the cortico-trigeminal connections in
the cat suggested by the following references: 1. Nakamura and
co-workers (1976), 2. Kidokoro and co-workers (1968a and b),
3. Bonvallet and Gary-Bobo (1975), 4. Rossi and Brodal (1956),
5. Kuypers (1958), 6. Wold and Brodal (1973), 7. Edwards (1972),
8. Landgren and Olsson (1977), 9. Brodal (1971), 10. Mizuno
and co-workers (1968), 11. Wold and Brodal (1974), 12. McBride
and Sutin (1977). Figures 1 - 12 are given in the diagram near
the relevant paths. Cx=cerebral cortex, LMC=lower motor centres,
NVmt=trigeminal motor nucleus, HVM=ventro medial hypothalamic
nucleus, DAA=defence attack area of the hypothalamus. Nret pont=
the pontine reticular nucleus, Nret gig=the gigantocellular
reticular nucleus, NsV=the supratrigeminal nucleus, NintV=the
intertrigeminal nucleus, NVsnpr=the main sensory trigeminal
nucleus, NVspo=the oral subdivision of the nucleus of the
spinal trigeminal tract, NVip=the interpolar nucleus of the
spinal trigeminal tract, Mass=masseteric motoneurone, Dig=
digastric motoneurone.

by the investigations of Kawamura and Tsukamoto (1960) and of Bonvallet and Gary-Bobo (1975). Anatomical evidence of the connections between amygdala and the Defence Attack Area of the hypothalamus were presented by McBride and Sutin (1977) as illustrated in Fig. 4. The course of the descending path from these nuclei is, however, an open question, alternative connections are therefore drawn with broken lines in the figure.

Effects on the jaw reflexes were evoked by electrical stimulation of the orbital gyrus (Fig. 2 E). Interesting evidence concerning the cortico-bulbar connections of this gyrus was presented by Brodal (1971), who demonstrated degenerating terminals in N pontis after lesions in the rostral part of the orbital gyrus. The caudal part of this gyrus projects to the interpolar nucleus of the spinal trigeminal tract as shown by Mizuno and co-workers (1968) and by Wold and Brodal (1974). The interpolar nucleus sends axons to the cerebellum (Darian-Smith and Phillips, 1964). The connections from the interpolar and the pontine nuclei to the trigeminal motor nucleus have so far not been investigated.

The diagram of Fig. 4 emphasizes, that many alternative paths between the cerebral cortex and the trigeminal motoneurones may exist. Most of them are not investigated in detail, and much experimentation is therefore needed to demonstrate their synaptic connections and their effects on the final common path. As quoted above, a precise knowledge of the interneurones in the lower motor centres adjacent to the trigminal motor nucleus is, however, developing. Some of these interneurones act in jaw reflexes, others may function in descending, ascending or proprio-trigeminal paths.

The path subserving the short latency facilitation and inhibition, which is described in this report, presumably correspond to the pyramidal or the cortico-rubro-trigeminal tracts shown to the left in Fig. 4. Furhter experiments are, however, required to determine their course and connections.

The observed sequence of cortically evoked facilitation and inhibition, alternating with a period of 10 ms, is much to fast to be related to normal chewing, which operates with 25 to 100 times longer periods. Our findings should therefore be interpreted as evidence of the capacity of the cerebral cortex to activate short latency excitatory and inhibitory paths to the masseteric and the digastric motoneurones and to select between them, thereby favouring jaw opening or jaw closing. These paths may be involved in long loop reflexes or in cortico-trigemino-coritcal feed back loops used in the correction of the motor command signals.

The existence of such mechanisms was suggested by Phillips (1969) and by Porter (1976). Their relevance for the control of jaw movements is substantiated by the observations of Luschei and Goodwin (1975), who demonstrated that monkeys, trained to produce a steady biting force on a strain gauge apparatus placed between their jaws, could not perform the force holding task after bilateral lesions of the precentral face area of the cerebral cortex. The experiments of Lund and Lamarre (1973) are also suggestive. They describe measurements of voluntary biting force in human subjects before and after local anaesthetic infiltration around the roots of the premolar teeth. The voluntary biting force was reduced by about 40% after the blocking of the sensory input from the teeth. The two sets of observations are compatible with the assumption of afferent reinforcement of the motor signals from the cerebral cortex, which could be subserved by fast long loop reflexes.

The role of the lower motor centres in the generation of the chewing rythm is another interesting problem. The existence of a central pattern generator for this rythm is genrally accepted (Lund, 1976a and b; Dubner, Sessle and Storey, 1978). Chewing may occur in the decerebrate cat, and can be evoked by electrical stimulation in the brain of a paralyzed animal lacking proprioceptive signals from the muscles. An automatic pattern generator is therefore assumed to be located in the medulla oblongata or the mesencephalon.

The location of the generator is still an open question. The best suggestions are those of Lund and Dellow (1971), who observed rythmic chewing in response to electrical stimulation in the reticular formation at the pontine level of the rabbit. It is difficult to locate origin of functional centres in this area rich in descending and ascending paths, and the correlation between function and structure is not yet precise. The location of two of their effective points in the mesencephalic reticular formation is, however, interesting, because they may correspond to nucleus reticularis pontis oralis, where Rossi and Brodal (1956) have described degenerating terminals after lesions involving the coronal gyrus of the cerebral cortex. The location of these points are also close to nucleus cuneiformis, which corresponds to the mesencephalic locomotor region of Shik and Orlovsky (1976).

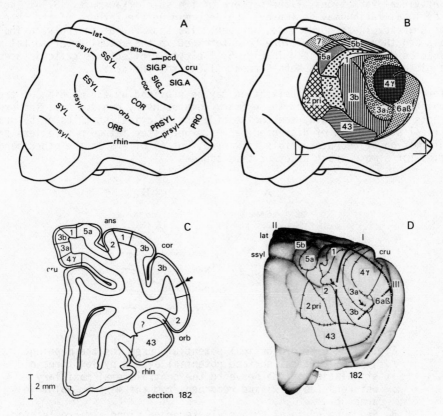

Fig. 5. Diagrams summarizing the method used for identifying cytoarchitectonic areas of the frontal part of the right hemisphere.
A. Nomenclature of sulci and gyri according to Hassler and Muhs-Clement (1964).
B. Cytoarchitectonic areas. The boundaries represent typical features observed in the individual experiments (cf. D).
C. Outline of a serial section (cf. sectioning plane I, section number 182 in D) with classified cytoarchitecture as determined according to the cirteria of Hassler and Muhs-Clement (1964).
D. Photograph of the perfused hemisphere from an individual experiment. Borders of the cytoarchitectonic areas (thin lines) are given.

The organization of the cerebral cortex controlling jaw movements

The cortical control of jaw movements may be studied by observing the animals behaviour in response to electrical stimulation of the cortex. Electromyograms from
the jaw muscles, compound action potentials from motor nerves, or postsynaptic potentials from the trigeminal motoneurones may be recorded. Cortical conditioning
of test reflexes, as used in the present report, may provide further evidence. In
all these types of experiments it is, however, necessary to describe the cortical
origin of the observed effects, if knowledge shall be gained concerning the organization of the cortical mechanisms.

Cortical localization of the site of stimulation may refer to the anatomical pattern
of sulci and gyri. Such landmarks are crude due to the dimensions of the gyri and to
the individual variations of the pattern of the sulci (Kawamura, 1971). The preci-
sion of localization may be improved by determining the borders of the cytoarchitectonic areas. These borders do, however, also show variations between individual
animals and they should therefore be determined in the actual experimental animal.
This can be done as described in Fig. 5. The cytoarchitectonic criteria used were
those given by Hassler and Muhs-Clement (1964) for the cat.

A reference to functional localization may be provided by the mapping of the maximum points and borderlines of the primary somatosensory projections. The projecting
afferents may be defined by the choice of a physiological stimulus or by graded
electrical stimulation of dissected nerves. The latter method may be less precise,
if information of receptor origin is wanted, but it offers on the other hand possibilities for a more exact study of time courses.

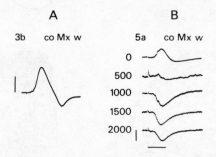

Fig. 6. Records of cortical potentials used for the mapping.
A. Typical cortical surface potential evoked by low threshold
   stimulation of the nerve to the contralateral maxillary
   whiskers (co Mx w) and recorded from the maximum point in
   area 3b.
B. Sequence of focal potentials recorded along a microelectrode
   track in the maximum point of area 5a. Calibration: 10 ms,
   500 μV. Recording depth in μm below the cortical surface.
   Positivity upwards.

The cortical signal recorded in response to electrical stimulation of the nerves is
the evoked potential. As shown in Fig. 6 A an initially positive wave is recorded
from the cortical surface. A characteristic reversal of polarity is observed, when
the potential is recorded along a microelectrode track penetrating perpendicularly
to the cortical surface. At depths corresponding to cortical layer IV and V the
initially negative phase of the evoked potential shows a maximal amplitude (Fig.
6 B). The phase reversal has been interpreted as due to termination of the thalamo-

cortical afferents on interneurones in cortical layer IV and the depolarization of pyramidal cells evoked by these interneurones. Based on this interpretation, the location of such evoked potentials gives the centre (maximum point) of the fastest input from the peripheral afferents to the cerebral cortex. Such a location may thus by defined as a primary projection centre.

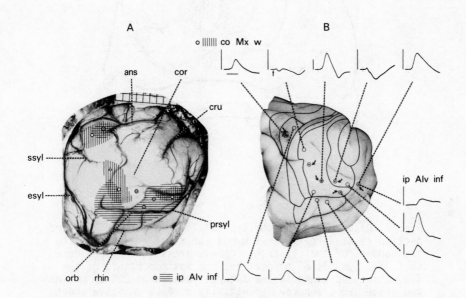

Fig. 7. Illustration of original data from one experiment.
A. Projection fields (horizontal hatching) and maximum points
   (⊙) of ip Alv inf. Vertical hatching: Fields of co Mx w.
   Maximum points of co Mx w:◯ and of Mx nose: △ . Maximum point of initially negative surface potentials evoked
   by co Mx w:⊖ . mm scale at the top of the photograph.
B. Records of evoked surface potentials, recording positions
   and cytoarchitectonic borders from the same experiment as
   in A. India ink punctures at the arrows. Time and voltage
   bars: 10 ms and 500 µV. Note differences in amplification.

We have mapped the maximum points and projection fields of ipsi- and contralateral low threshold afferents from the oral cavity and the face of the cat. The cytoarchitectonic borders were determined in the actual experimental animal (Landgren and Olsson, 1980). Original data from such an experiment are shown in Fig. 7. Separately located maximum points and projection fields were observed in several cytoarchitectonic areas. A standardized map summarizing typical results from our series of experiments is displayed in Fig. 8. The projections from the low threshold afferents of the nerve to the contralateral maxillary whiskers (co Mx w) were selected for presentation. These afferents show primary projection centres in areas 5a, 3b, 3a and 6aβ. A primary projection to area 4γ was assumed to exist, but was not localized due to the fact that a reversal of the surface positive evoked potential recorded from area 4γ was not found.

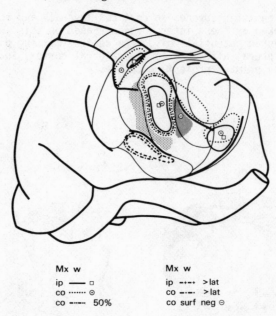

Mx w   Mx w

ip —— □   ip –·–· > lat

co ······· ⊙   co –··– > lat

co –··– 50%   co surf neg ⊖

Fig. 8. Diagram showing the typical location of the Mx w projection related to the anatomy of the sulci and gyri and to
the borders of the cytoarchitectonic areas (thin lines drawn
in full). Symbols of borderlines and maximum points of the
fields are explained in the figure. Borders at 20% of maximal
amplitude are given for the initially surface positive short
latency potentials unless otherwise indicated. Dense stippling:
field of initally negative or diphasic positive-negative
potentials, maximum point: ⊖ . Thin stippling: field of
diphasic negative-positive potentials.

We interpret the maximum points as centres of afferent input to functionally
different cortical areas. This interpretation was tested by an investigation of
the effects of cortical stimulation on the monosynaptic jaw closing and the di-
synatpic jaw opening reflexes. The cortical points selected for stimulation were
localized in relation to the pattern of the gyri and sulci, the borders of the
cytoarchitectonic areas, and the maximum points of the projection fields of low
threshold afferents from the oral cavity and the face.

As shown in Fig. 9 the fastest, largest, and most complex effects on the jaw
reflexes were evoked from the projections of the whiskers and the teeth to area
3a of the coronal gyrus (cf. Fig. 9 C and E). The maximum point of the whisker
projection to area 3b also evoked large responses. Rather unexpectedly, the motor
cortex proper, i.e. area 4γ was less effective as seen from Fig. 9 G, and so were
the maximum points in areas 5a and 6aβ.

The observation that powerful effects on the jaw reflexes could be evoked from
area 3a is interesting. The afferents from whiskers and periodontal ligaments are
obviously activated during the movements of the face and the jaws. An input related to movements was also demonstrated in area 3a of the posterior sigmoid gyrus.
Low threshold muscle afferents from the contralateral fore- and hindlimbs thus

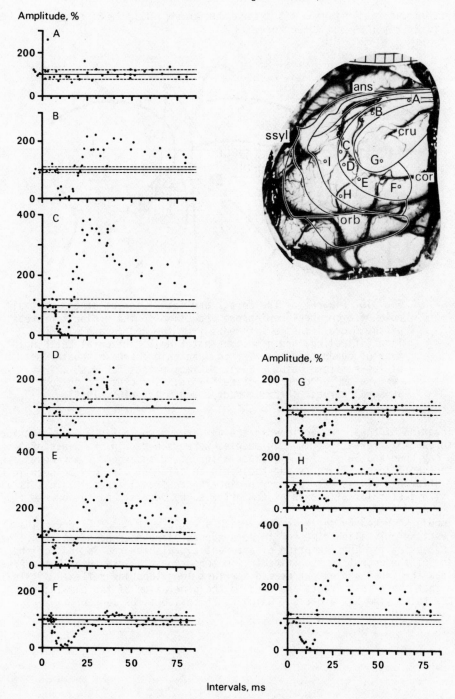

Fig. 9. Effects of a cortical conditioning train (anode, 3 pulses, 0.5 ms, 555 Hz, 0.6 mA) on the monosynaptic jaw closing reflex. The locations of the stimuli in relation to the defined cytoarchitectonic areas given in the photograph of the brain of the experimental animal. A-F located in area 3a, G in 4γ and H, I in 3b.

project to the latter region (Oscarsson and Rosén, 1963, 1966; Landgren and Silfve nius, 1969; Phillips and co-workers, 1971).

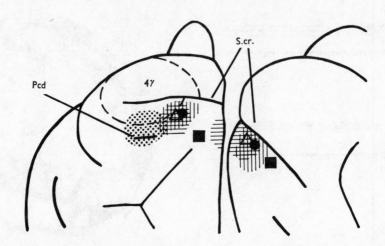

Fig. 10. Diagram of the dorsal and medial aspect of the rostral pole of a cerebral hemisphere. Location of the projection areas of the Group I muscle afferents from the contralateral fore-limb (stippling) and hindlimb (hatching). Horizontal hatching: area of quadriceps (Q). Vertical hatching: area of posterior biceps-semitendinosus (PBSt). Maximum points for Q: △ , PBSt: ● , Su: ■ , Pcd=postcruciate dimple. S.cr=cruciate sulcus, Su=sural nerves. (From: Landgren and Silfvenius, 1969).

The location of these projections in the cat are shown in Fig. 10. Projections from low threshold afferents of the jaw muscles were recorded in the banks of the coron sulcus by Lund and Sessle (1974). Other afferents discharged by movements are the low threshold joint afferents and the Pacinian afferents. As may be seen in Fig. 1 these afferents also showed primary projection centres in area 3a. This is also the case with vestibular afferents as demonstrated by Ödqvist and co-workers (1975).

The impulses evoked by the movements reach area 3a with short latency. The shortes observations only allow time for three synapses along the ascending path. The time necessary for the reception of afferent signals in area 3a and for the pro-ceeding of the output to the trigeminal motoneurones is less than 10 ms. Area 3a thus has the input and output needed for fast long loop jaw reflexes. It is impro-bable that these fast events are used in the generation of the chewing rythm, because this rythm has a period, which is 25 to 100 times longer.

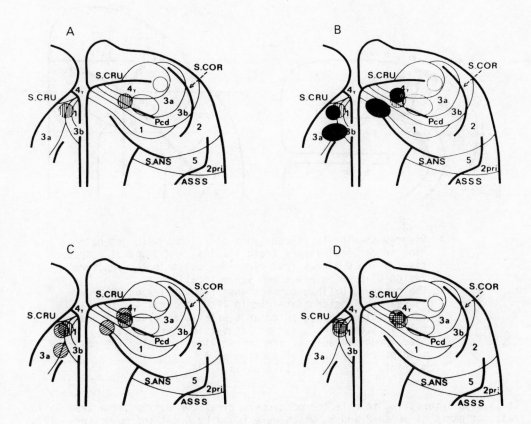

Fig. 11. Diagrams showing the projections to the cerebral cortex
from proprioceptive and exteroceptive afferents.
A. Projections of group I muscle afferents from the contralateral hindlimb (vertical hatching) and forelimb (encircled).
   The group I projections to the lateral ansate and anterior
   suprasylvian sulci are not shown.
B. Low threshold skin afferents (sural nerve, black) and group I
   muscle afferents.
C. Low threshold joint afferents (contralateral posterior Knee
   joint nerve, obligue hatching) and group I muscle afferents.
D. Pacinian afferents (contralateral interosseus nerve, horizontal hatching) and group I muscle afferents.
Borders of the cytoarchitectonic areas according to Hassler
and Muhs-Clement (1964). From Silfvenius, 1972.

Cortico-cortical U-fibre connections from area 3a were described by Grant and co-
workers (1975). As illustrated in Fig. 12 such U-fibre connections showed a columnar termination and were distributed to the forelimb and hindlimb subdivisions of
area 4γ. Zarzecki, Shinoda and Asanuma (1978) have confirmed the observation of
this U-fibre link, and showed that excitation and inhibition of the link could be
evoked by stimulation of group I muscle afferents.

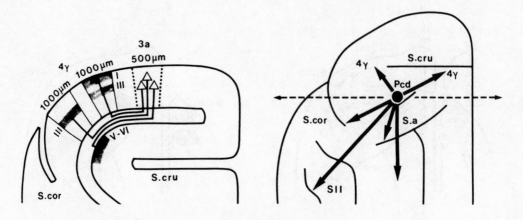

Fig. 12.
A. Diagram showing the three types of columns with degeneration in cortical layers I and III,III, and V and VI respectively. The U-fibres from pyramidal cells in the puncture
   lesion of area 3a are drawn schematically.
B. The observed U-fibre connections from the lesion in area
   3a near the postcruciate dimple (Pcd) to area 4γ, lateral
   ansate sulcus (S.a), anterior suprasylvian gyrus, coronal
   sulcus (S.cor) and anterior ectosylvian gyrus (SII).
   S.cru=cruciate sulcus. The broken line indicates the position of the transversal section of A.
From Grant, Landgren and Silfvenius, 1976.

It is also interesting to note the cortico-cortical connections from area 3a to
adjacent parts of areas 2 and 5, which were found by Grant and co-workers (1975).
Area 2 also receives proprioceptive projections (Silfvenius, 1968, 1972), and initiation of motivated arm movements from area 5 was inferred from the experiments
of Mountcastle and co-workers (1975). A cortico-cortical link from area 5 back to
the motor area 4 was further described by Zarzecki, Strick and Asanuma (1978).

All these observations emphasize the engagement of area 3a in cortico-cortical
connections related to motor functions. It is thus likley that this area is not
only a centre for long loop reflexes but also a component integrated in complex
mechanisms like those controlling the jaws in the process of catching the prey.

ACKNOWLEDGEMENT

This work was supported by the Swedish Medical Research Council (Proj. B79-14X-00045-
15) and by Gunvor and Josef Anérs Stiftelse.

REFERENCES

Bonvallet, M., and E. Gary-Bobo (1975). Amygdala and masseteric reflex. II. Mechanism of the disynaptic modifications of the reflex elicited from the "defence reaction area". Role of the spinal trigeminal nucleus (pars oralis). Electroenceph. clin. Neurophysiol., 21, 209-226.

Brodal, P. (1971). The corticopontine projection in the cat. II. The projection from the orbital gyrus. J. comp. Neurol., 142, 141-152.

Chase, M.H., M.B. Sterman, K. Kubota, C.D. Clemente (1973). Modulation of masseteric and digastric neural activity by stimulation of the dorsolateral cerebral cortex in the squirrel monkey. Exp. Neurol, 41, 277-289.

Darian-Smith, I., and G Phillips (1964). Secondary neurones within a trigeminocerebellar projection to the anterior lobe of the cerebellum in the cat. J. Physiol. (Lond.), 170, 53-68.

Dubner, R., B.J. Sessle, and A.T. Storey (1978). The neural basis of oral and facial function. Plenum Press, New York.

Dobrovsky, B., E. Garcia-Rill, and M.A. Surkes (1974). Effects of discrete precruciate cortex lesions on motor behaviour. Brain Res., 82, 328-333.

Edwards, S.B. (1972). The ascending and descending projections of the red nucleus in the cat: an experimental study using an autoradiographic tracing method. Brain Res., 48, 45-63.

Grant, G., S. Landgren, and H. Silfvenius (1975). Columnar distribution of U-fibres from the postcruciate cerebral projection area of the cat's group I muscle afferents. Exp. Brain Res., 24, 57-74.

Grant, G., S. Landgren, and H. Silfvenius (1976). Columnar distribution of U-fibres from the postcruciate cerebral projection area of the cat's group I muscle afferents. Exp. Brain Res., Suppl. 1., pp. 317-322.

Hassler, R., and K. Muhs-Clement (1964). Architektonischer Aufbau des sensomotrischen und parietalen Cortex der Katze. J. Hirnforsch., 6, 377-420.

Hess, W.R., K. Akert, and D.A. McDonald (1952). Functions of the orbital gyri of cats. Brain, 75, 244-259.

Kawamura, K. (1971). Variations of the cerebral sulci in the cat. Acta Anat., 80, 204-221.

Kawamura, Y., and S. Tsukamoto (1960). Analysis of jaw movement from the cortical jaw motor area and amygdala. Jap. J. Physiol., 10, 471-488.

Kidokoro, Y., K. Kubota, S. Shuto, and R. Sumino (1968a). Reflex organization of cat masticatory muscles. J. Neurophysiol, 31, 695-708.

Kidokoro, Y., K. Kubota, S. Shuto, and R. Sumino (1968b). Possible interneurones responsible for reflex inhibition of motoneurones of jaw closing muscles from the inferior dental nerve. J. Neurophysiol., 31, 709-716.

Kubota, K. (1976). Motoneurone mechanisms: suprasegmental controls. In: B.J. Sessle, A.G. Hannam (Eds.), Mastication and Swallowing. University Toronto Press, Toronto, pp. 60-75.

Kuypers, H.G.J.M. (1958). An anatomical analysis of corticobulbar connections to the pons and lower brain stem in the cat. J. Anat., 92, 198-218.

Landgren, S., and H. Silfvenius (1969). Projection to cerebral cortex of group I muscle afferents from the cat's hind limb. J. Physiol. (Lond.), 200, 353-372.

Landgren, S., and K.A. Olsson (1977). The effect of electrical stimulation in the defence attack area of the hypothalamus on the monosynaptic jaw closing and disynaptic jaw opening reflexes in the cat. In: D.J. Anderson and B. Matthews (Eds.), Pain in the trigeminal region. Elsevier/North-Holland Biomedical Press, Amsterdam. pp. 385-394.

Landgren, S., and K.A. Olsson (1980). Low threshold afferent projections from the oral cavity and the face to the cerebral cortex of the cat. (To be published).

Lund, J.P., and P.G. Dellow (1971). The influence of interactive stimuli on rythmical masticatory movements in rabbits. Archs. oral. Biol., 16, 215-223.

Lund, J.P. (1976a). Evidence for a central neural pattern generator regulating the chewing cycle. In D.J. Anderson and B. Matthews (Eds.), Mastication. pp. 204-212.

Lund, J.P. (1976b). Oral-facial sensation in the control of mastication and volun-
    tary movements of the jaw. In B.J. Sessle and A.G. Hannam (Eds.), Mastication
    and Swallowing. Univ. Toronto Press. Toronto.
Lund, J.P., and Y. Lamarre (1973). The importance of positive feedback from perio-
    dontal pressoreceptors during voluntary isometric contraction of jaw closing
    muscles in man. J. Biol. Buccale., 1, 345-351.
Lund, J.P., and B.J. Sessle (1974). Oral-facial and jaw muscle afferent projections
    to neurones in cat frontal cortex. Exp. Neurol., 45, 314-331.
Lundberg, A. (1969). Reflex control of stepping. The Norwegian Academy of Science
    and Letters. Universitetsforlaget, Oslo. pp. 5-42.
Luschei, E.S., and G.M. Goodwin (1975). Role of monkey precentral cortex in control
    of voluntary jaw movements. J. Neurophysiol., 38, 146-157.
Magoun, H.W., S.W. Ranson, and C. Fischer (1933). Corticofugal pathways for masti-
    cation, lapping and other motor functions in the cat. Archs. Neurol. Psychiat.,
    30, 292-308.
McBride, R.L., and J. Sutin (1977). Amygdaloid and pontine projections to the
    ventromedial nucleus of the hypothalamus. J. comp. Neurol., 174, 377-396.
Mizuno, N., E.K. Sauerland, C.D. Clemente (1968). Projection from the orbital gyrus
    in the cat. I. To brain stem structures. J. comp. Neurol., 133, 463-475.
Morimoto, T., and Y. Kawamura (1973). Properties of tongue and jaw movements elici-
    ted by stimulation of the orbital gyrus in the cat. Archs. oral Biol., 18,
    361-372.
Mountcastle, V.B., J.C. Lynch, A. Georgopoulus, H. Sakata, and A. Acuna (1975).
    Posterior parietal association cortex of the monkey: Command functions for
    operations within extra personal space. J. Neurophysiol., 38, 871-908.
Nakamura, Y., Y. Kubo, S. Nozaki, and M. Takatori (1976). Cortically induced masti-
    catory rythm and it's modification by tonic peripheral inputs in immobilized
    cats. Bull. Tokyo Med. Dent. Univ., 23, 101-107.
Nakamura, Y., M. Takatori, S. Nozaki, and M. Kikuchi (1975). Monosynaptic reci-
    procal control of trigeminal motoneurones from the medial bulbar reticular
    formation. Brain Res., 89, 144-148.
Nakamura, Y., S. Nozaki, M. Takatori, and M. Kikuchi (1976). Possible inhibitory
    neurones in the bulbar reticular formation involved in the cortically evoked
    inhibition of the masseteric motoneurone of the cat. Brain Res., 115, 512-517.
Ödqvist, L., B. Larsby, and J.M. Fredrickson (1975). Projection of the vestibular
    nerve to the SI arm field in the cerebral cortex of the cat. Acta oto-laryng.
    (Stockh.), 79, 88-95.
Olsson, K.Å., and S. Landgren (1980). Facilitation and inhibition of jaw reflexes
    evoked by electrical stimulation of the cat's cerebral cortex. (To be published)
Oscarsson, O., and I. Rosén (1963). Projection to cerebral cortex of large muscle
    spindle afferents in forelimb nerves of the cat. J. Physiol. (Lond.), 169,
    924-945.
Oscarsson, O., and I. Rosén (1966). Short-latency projections to the cat's cerebral
    cortex from skin and muscle afferents in the contralateral forelimb. J. Physiol.
    (Lond.), 182, 164-184.
Phillips, C.G. (1969). Motor apparatus of the baboon's hand. Proc. R. Soc. B.,
    173, 141-174.
Phillips, C.G., T.P.S. Powell, and M. Wiesendauger (1971). Projection from low
    threshold muscle afferents of hand and forearm to area 3a of baboon's cortex.
    J. Physiol. (Lond.), 217, 419-446.
Porter, R. (1976). Influences of movement detectors on pyramidal tract neurones
    in primates. Ann. Rev. Physiol, 38, 121-137.
Rossi, G.F. and A. Brodal (1956). Corticofugal fibres to the brain-stem reticular
    formation. An experimental study in the cat. J. Anat., 90, 42-62.
Sauerland, E.K., Y. Nakamura, C.D. Clemente (1967). The role of the lower brain
    stem in cortically induced inhibition of somatic reflexes in the cat. Brain
    Res., 6, 164-180.
Shik, M.L., and G.N. Orlovsky (1976). Neurophysiology of locomotor automatism.

Physiol. Reviews, 56, 464-501.
Silfvenius, H. (1968). Cortical projections of large muscle spindle afferents from the cats's forelimb. Acta Physiol. Scand., 74, 25-26A.
Silfvenius, H. (1972). Projection to the cat cerebral cortex from fore- and hind-limb group I muscle afferents. Umeå Univ. Med. Diss. No 4.
Wahlberg, F. (1957). Do the motor nuclei of the cranial nerves receive cortico-fugal fibres? An experimental study in the cat. Brain, 80, 597-605.
Wold, J.E., and A. Brodal (1973). The projection of cortical sensimotor regions onto the trigeminal nucleus in the cat. An experimental anatomical study. Neurobiol., 3, 353-375.
Wold, J.E., and A. Brodal (1974). The cortical projection of the orbital and proreate gyri to the sensory trigeminal nuclei in the cat. An experimental anatomical study. Brain Res., 65, 381-395.
Zarzecki, P., Y. Shinoda, and H. Asanuma (1978). Projection from area 3a to the motor cortex by neurones activated from group I muscle afferents. Exp. Brain Res., 33, 269-282.
Zarzecki, P., P.L. Strick, and H. Asanuma (1978). Input to primate motor cortex from posterior parietal cortex (area 5). II Identification by antidromic acti-vation. Brain Res., 157, 331-335.

THERMORECEPTION

H. Hensel

Institute of Physiology, University of Marburg, Deutschhausstraße 2,
D-3550 Marburg, Federal Republic of Germany

ABSTRACT

Thermoreception includes temperature sensation, thermal comfort, temperature regulation, and thermal behavior. A dual neural system. underlies cold and warm reception. Dynamic thermal sensations during rapid temperature changes correspond with transient overshoots in thermoreceptor activity. Static cold sensations at lower temperatures can neither be correlated with the average discharge nor with the burst parameters of cold fibers. The neural correlate of these cold sensations as well as of cold pain is possibly an across-fiber pattern including low-temperature receptors. Human warm receptors show a dynamic overshoot on rapid warming and a static activity with maxima between $41^{\circ}C$ and $45^{\circ}C$, which corresponds well with warm sensation. The burst discharge of cold receptors is assumed to be triggered by an oscillating receptor potential due to a temperature-dependent Na^+/K^+ permeability ratio and a negative feedback from a temperature- and potential-dependent calcium-potassium system. Decrease in activity at higher temperatures is ascribed to hyperpolarization by a temperature-dependent electrogenic $Na^+ K^+$ pump. Ca^{++} decreases the mean static frequency of cold receptors and transforms bursts into a regular discharge, while EDTA increases the mean frequency and transforms a regular sequence of impulses into a burst discharge.

KEYWORDS

Thermoreception; cold receptors; warm receptors; across-fiber pattern; burst discharge; calcium; EDTA.

INTRODUCTION

The term 'Thermoreception' was introduced into Physiology in 1952
and adopted by the Encyclopaedia Britannica in 1974, defining thermoreception as "a process in which different levels of heat energy
(temperatures) are detected by living things". (Hensel, 1974). Thermoreception is involved in temperature sensation, thermal comfort,
temperature regulation, and thermal behavior (Fig. 1). Whereas tem-

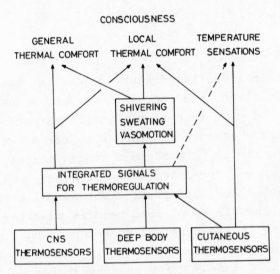

Fig. 1.   Schematic diagram of sensory inputs for
          temperature sensation, thermal comfort,
          and temperature regulation. From Hensel
          (1977).

perature sensation seems to be correlated merely with the activity
of cutaneous thermoreceptors and a direct pathway to the cerebral
cortex, thermal comfort, temperature regulation, and thermal behavior depend on the activity of both external and internal thermosen-
sors and the integration of their signals in the central nervous
system.

In the following, I would like to deal with some problems of cold
reception, warm reception, and transducer mechanisms of thermoreceptors, the results and speculations including contributions of my co-

workers H. Bade, R. Beste, H. A. Braun, F. Konietzny, and K. Schäfer.
It is now well established that the sensations of warm and cold are
mediated by two different neural systems, with specific warm and cold
receptors, separate afferent fiber systems and possibly some separation in the central pathways. It is also well known that the transient warm and cold sensations elicited by rapid temperature changes
are correlated with a dynamic overshoot in the activity of warm and
cold receptors, respectively (for references see Hensel, 1973, 1976).

STATIC COLD RECEPTION

When Zotterman and I discovered in 1950 that cold receptors show a
steady discharge at constant temperatures (Hensel and Zotterman,
1951), nobody would believe us, because this contradicted Weber's
classical theory of temperature sensation, according to which only
temperature changes are the adequate stimulus. Today the static discharge is a well known phenomenon of warm and cold receptors in a
variety of species including man. The average static discharge frequency of larger populations of cold receptors in the cat's nose has
a maximum at about $28^{\circ}C$, and the maximum for a population of cold
fibers in the monkey's hand is at $24^{\circ}C$ (Dykes, 1975). Warm receptors
in monkey and man start discharging at about $30^{\circ}C$ and reach their
static maxima at temperatures between $41^{\circ}C$ and $46^{\circ}C$.

The static discharge of cold receptors induced us to investigate
more thoroughly the static temperature sensation in human subjects
(Beste and Hensel, 1977; Beste, 1977). Without going into details,
the general result for the hand of 18 subjects is shown in Fig. 2.
At constant temperatures below $34^{\circ}C$, there is an increasing cold
sensation and a good discrimination of various static temperatures
until a temperature of about $27^{\circ}C$ is reached. There the curve becomes horizontal which means that no static temperature discrimination is possible in this range. But below $22^{\circ}C$, the cold sensation
increases again, and the discrimination is improved. Below $20^{\circ}C$, the
cold sensation changes not only in intensity but also in quality, in
that it becomes more and more painful.

If we compare the static cold sensation with the average discharge
frequency of a cold fiber population in the monkey's hand - the data
from the human hand are not sufficient as yet -, there is a fairly
good correlation down to about $24^{\circ}C$. But below this level, the static

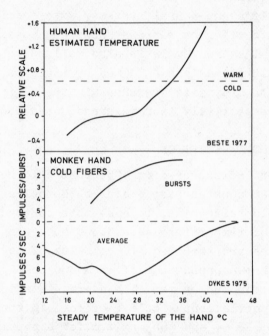

Fig. 2.   Estimated static temperatures in humans
          (average of 18 subjects) and static dis-
          charge parameters of a cold receptor
          population in the monkey at various con-
          stant temperatures of the hand. From
          Hensel (1977).

cold sensation increases, whereas the average frequency decreases.

In the range between 36°C and 20°C, periodic bursts of impulses in-
terrupted by silent intervals occur, the parameters of which change
monotonically with temperature. For instance, the number of impulses
per burst increases with decreasing temperature down to 20°C. But
neither can the burst discharge in the monkey account for the sensory phenomena in humans. For example, at 25°C, a steep slope of the
burst parameters as a function of temperature coincides with a zero
slope of cold sensation. Below 20°C, the burst discharge ceases but
the cold sensation increases particularly in this range.

A further argument against the burst discharge as an information
parameter for low constant temperatures may be derived from the
study of neurons in the thermosensitive pathway. We have no evidence

that the bursts are directly or indirectly conveyed to higher-order
neurons, their temperature-frequency curve being similar to the bell-
shaped curve of peripheral cold receptors (Poulos, 1975; Iggo and
Ramsey, 1976; Schmidt, 1976; Dostrovsky and Hellon, 1978).

Finally, it is doubtful whether the majority of human cold receptors
is bursting at all at constant temperatures, in contrast to the monkey, where this phenomenon occurs so regularly that the burst discharge can be used for the identification of a specific cold fiber
(Iggo, 1969, 1970). Neither in previous experiments (Hensel and
Boman, 1960) on the human hand did we find cold fibers with distinct
static bursting, nor did we see burst discharges in recent recordings with microelectrodes from the same area (Fig. 3).

Fig. 3.   Discharge of a single cold fiber in hu-
          man hairy skin. A, cooling from 32 to
          $30^\circ C$; B, from 32 to $25^\circ C$. From Konietzny
          and Hensel (1979, unpublished).

ACROSS-FIBER PATTERN

Therefore it seems worthwhile to look for new candidates as neurophysiological correlates of cold sensations in the low temperature
range. As has been mentioned already by Erickson and Poulos (1973),
across-fiber patterns should be considered in connection with the
phenomena of static temperature sensation. This would mean that the
sensation at different temperature levels may not only be a matter
of intensity but also of quality.

In connection with recent experiments on long-term thermal adaptation (Hensel and Schäfer, 1979), we have collected a population of
200 single cold fibers from the cat's nose using well defined static

and dynamic thermal stimuli between 10°C and 40°C. Each cold fiber
was carefully tested for specificity. The majority had static maxima
at 25°C to 35°C, but there were also cold fibers having maxima at
20°C, 15°C and 10°C, and a few fibers had maxima at 40°C. There were
some differences in cats adapted to 30°C and to 5°C ambient temperature that will not be discussed here. In Fig. 4, the cold fibers are

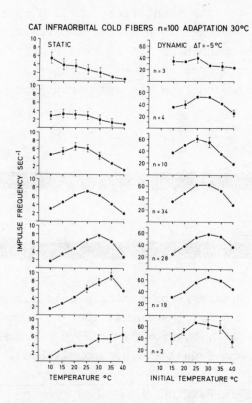

Fig. 4. Average static discharge frequency and
maximal dynamic response of single cold
fiber populations from the cat nose as
function of temperature. Dynamic responses were obtained by cooling steps of
5°C. Groups according to static maxima
at 10, 15, 20, 25, 30, 35 and 40°C, n =
number of fibers in each group. Cats
were adapted to ambients of 30°C for 4
years. Bars indicate SEM. From Hensel
and Schäfer (1979, unpublished).

comprised in groups according to their static maxima, each maximum
being 5°C lower than the previous one. For each group, the average
frequency is plotted as function of temperature. As can be seen, the
static maxima of cold fibers in the cat's nose cover the whole temperature range between 10°C and 40°C, while the dynamic maxima to
5°C cooling steps are concentrated around initial temperatures between 25°C and 30°C.

From these data I have tried to draw an across-fiber pattern at various constant temperatures (Fig. 5). For 7 groups classified by their

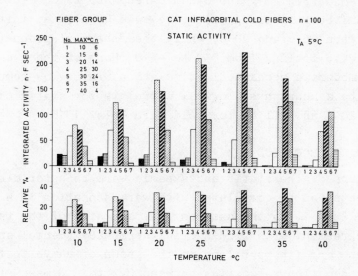

Fig. 5. Across-fiber pattern of static activity
         for 7 groups of cold fibers from the cat
         nose at various constant temperatures.
         Groups are indicated in the inset. Upper
         graph: integrated activity (average fre-
         quency multiplied with number of fibers
         for each group). Lower graph: relative
         integrated activity in percent. Cats
         were adapted to ambients of 5°C for 4
         years. From Hensel and Schäfer (1979,
         unpublished).

static maxima between 10°C and 40°C, the integrated frequency, that
is, the average frequency multiplied with the number of fibers, is
depicted for temperatures between 10°C and 40°C. The upper part of

Fig. 5 shows absolute values, the lower part relative values. Obviously there is a considerable change of the across-fiber pattern with temperature, in particular in the lower range from $30^{\circ}C$ to $10^{\circ}C$.

LOW-TEMPERATURE RECEPTORS

Recently we have studied the static and dynamic activity of cold-sensitive receptors in the glabrous skin of the cat's nose at skin temperatures as low as $-5^{\circ}C$ (Duclaux, Schäfer and Hensel, 1979). We found a number of fibers with static maxima at $10^{\circ}C$, $5^{\circ}C$, and 2 fibers even increased their frequency when the temperature was lowered from 15 to $-5^{\circ}C$ (Fig. 6). The dynamic response to cooling steps of $5^{\circ}C$ for the fibers in group (a) to (e) was highest at initial temperatures between $25^{\circ}C$ and $20^{\circ}C$, while the fibers of group (f) responded maximally to cooling from $0^{\circ}C$ to $-5^{\circ}C$. None of the receptors did respond to moderate mechanical stimulation. These low-temperature fibers may be a neurophysiological correlate for static cold sensation below $20^{\circ}C$ as well as for cold pain below $10^{\circ}C$.

WARM RECEPTION

The parameters of the static and dynamic warm receptor discharge in human subjects can be correlated with warm sensation in a more satisfactory way than it is the case on the cold side of the temperature range. In agreement with findings in the monkey (Hensel and Iggo, 1971), human warm fibers belong to the group of unmyelinated C fibers with conduction velocities between 0.5 to 0.8 $msec^{-1}$ (Konietzny and Hensel, 1975). Fig. 7 A shows the static discharge of a warm fiber in the human hand at various temperatures, Fig. 7 B the impulse frequency as function of constant temperature. The receptor shows a dynamic sensitivity to temperature rises, the transient overshoot in frequency being a function of the slope of the linear dynamic warm stimulus (Fig. 7 C).

Starting from one and the same adapting temperature, the maximal dynamic response is linearly related to the magnitude of the dynamic temperature increment (Fig. 8 A). For a given increment, the dynamic response increases also with the level of the adapting temperature (Fig. 8 B). Thus the relationship between temperature increment and maximal dynamic response becomes steeper with higher levels of the adapting temperature (Fig. 8 C). These properties of human warm fi-

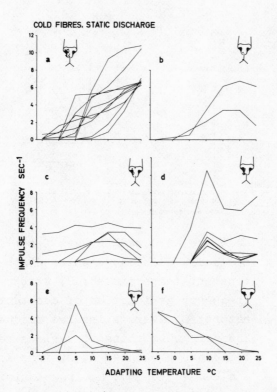

COLD FIBRES. STATIC DISCHARGE

Fig. 6.   Static impulse frequency of 27 single
           cold fibers from the cat nose as function
           of constant temperatures. Fibers com-
           prised in groups according to static
           maxima as follows: a, 25°C or above;
           b, 20°C; c, 15°C; d, 10°C; e, 5°C; f,
           -5°C or below. Insets: receptive fields
           in the glabrous skin of the nose. From
           Duclaux, Schäfer and Hensel (1979, un-
           published).

bers correspond well with the facts of dynamic warm sensation.

The same is also true for the static warm sensation. In the range between 34°C and 45°C, the intensity of warm sensation on the human
hand increases in a similar way (cf. Fig. 2) as does the average
static impulse frequency of warm receptors in the same area (Fig. 9).
The static maxima of warm receptors in the human hand vary between
41°C and 45°C; some receptors may have maxima at even higher temperatures, but in our subjects we did not exceed 46°C in order to avoid

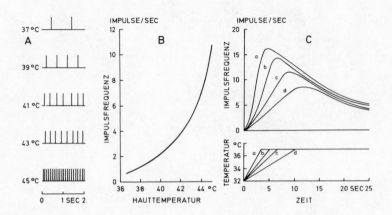

Fig. 7.   Response of a single warm fiber from hu-
           man hairy skin. A, impulse discharge at
           constant temperatures; B, static impulse
           frequency as function of constant tempe-
           rature; C, average impulse frequencies of
           2 single warm units at various rates of
           temperature change. A, B, from Konietzny
           and Hensel (1975); C, from Konietzny and
           Hensel (1977).

excessive heat pain and damage of the skin. In the monkey's foot
(Hensel, 1973), the distribution of the static maxima of warm fibers
is very similar to that found for humans, the only difference being
a lower static frequency of warm fibers in the human hand.

                        TRANSDUCER MECHANISMS

How can static temperatures and dynamic temperature changes be transformed into afferent impulses? Since thermoreceptors are not accessible to intracellular recording with microelectrodes, we are to rely on more indirect methods, such as analysing impulse discharge patterns under various thermal and chemical conditions.

In order to describe the bell-shaped frequency-temperature curve of
the static discharge as well as the transient overshoot and under-
shoot in frequency on dynamic temperature changes, more than one temperature-dependent process must be assumed. In 1938, Sand proposed a
model for the response of the ampullae of Lorenzini to thermal stimuli. He postulated two processes with positive temperature coeffi-

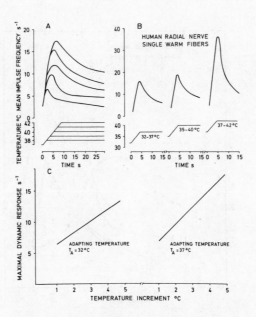

Fig. 8.   A, average impulse frequencies of a single warm fiber from human hairy skin when
applying linear temperature changes of
constant rate and various magnitude at
37°C adapting temperature. B, impulse
frequencies of a single warm unit and
skin temperature when applying linear
temperature changes of constant rate and
equal magnitude at 32, 35 and 37°C adapting temperature. C, maximal dynamic response at adapting temperatures of 32°C
and 37°C. From Konietzny and Hensel(1977).

cients, one process being excitatory, the other inhibitory. The actual response would correspond to the difference between both processes.
This model was purely formal in nature, and so were several other
models for thermoreceptors, models that can be considered as modifications of Sand's classical concept.

Much has been learned in recent years from pacemaker neurons in molluscs, such as Aplysia and Helix (Barker and Gainer, 1975; Meech and
Standen, 1975; Johnston, 1976; Gola, 1976; Eckert and Lux, 1976).
These neurons show a temperature-dependent regular or bursting steady
discharge and have been thoroughly studied with intracellular record-

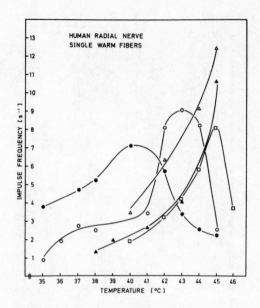

Fig. 9.   Static discharge frequency of single
          warm fibers from human hairy skin of the
          hand as function of constant temperature.
          From Konietzny and Hensel (1979, unpub-
          lished).

ings, measurements of ion fluxes under various conditions, and application of inhibitors. These findings, in combination with recordings
of cold receptor discharges under ouabain and various extracellular
ion concentrations led to the hypothesis that the inhibitory process
of the cold receptor may be the activity of an electrogenic sodium
pump having a positive temperature coefficient and thus leading to
hyperpolarization of the receptor membrane with increasing temperature. The excitatory process was assumed to be the Na^+/K^+ permeability ratio which increases also with temperature and thus leads to increasing depolarization of the receptor membrane. The static and dynamic impulse frequencies were assumed to depend on the difference
between the two processes (Pierau, Torrey and Carpenter, 1975).

In this connection, the burst discharge of cold receptors became
highly interesting as a clue to the receptor mechanisms. From intracellular recordings of bursting neurons in molluscs we may assume
that the bursts in a cold receptor are triggered by an oscillating
receptor potential when it exceeds a threshold of depolarization.

The frequency of bursts will depend on the frequency of the oscillation, while the intraburst frequency will increase with the amount
of depolarization during each phase of the oscillation. At a given
frequency of the oscillation, an increasing depolarization will also
lead to a higher number of spikes per burst.

By detailed analysis of the time intervals of impulses within the
burst we have tried to reconstruct the shape of the hypothetical oscillation for 3 different temperatures (Fig. 10). With rising tempe-

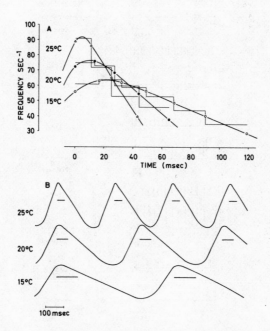

Fig. 10.   Hypothetical burst-triggering oscilla-
           tions in a lingual cold receptor in the
           cat. A, curves representing possible
           burst-triggering phases of oscillating
           receptor potential. They are drawn from
           the mean frequency-time histograms
           (steps) of the intraburst discharge.
           Triangles, points, and circles indicate
           occurrence of spikes. B, hypothetical
           oscillations; bars indicate burst dura-
           tion. From Braun, Bade and Hensel (1979).

rature, the oscillation increases in frequency and amplitude, there-

by leading to a higher frequency of bursts and a higher intraburst
frequency. Because of the shorter peaks of the oscillation, the burs
duration and thus the number of spikes per burst decreases.

Of particular interest are the events at higher constant temperature
In the example in Fig. 11 A, the bursts appear in regular doublets a

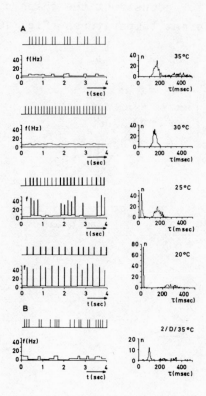

Fig. 11.   Discharge pattern, instantaneous frequen-
           cy, and interval histogram of lingual
           cold receptors in the cat. A, record
           from a cold receptor showing continuous
           transition from regular bursting at 20°C
           to irregular spike discharge at 35°C. B,
           record from another cold receptor. From
           Braun, Bade and Hensel (1979).

20°C. At 25°C, an irregular alternation of doublets and single spikes
occurs. As can be seen in the interval histogram, the time interval
between the single spikes is the same as that between the bursts. At

30°C, a regular discharge of single spikes occurs, the intervals being but slightly shorter than the longer intervals at 30°C. Finally,
at 35°C, the discharge becomes irregular. The longer intervals are
not randomly distributed, but are multiples of the shortest intervals.
This pattern is even more pronounced in another cold fiber (Fig. 11B).
We can therefore conclude that the oscillation of the receptor potential is maintained throughout, and even beyond, the whole range of
static activity, its frequency and amplitude increasing with rising
temperature.

At higher temperatures, however, an increasing hyperpolarization of
the resting potential occurs, due to the activity of a temperature-
dependent electrogenic $Na^+ K^+$ pump, so that the peaks of the oscillation finally fail to reach the threshold for the initiation of impulses (Fig. 12).

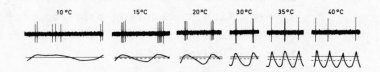

Fig. 12.   Discharge of a single cold fiber from
           the lingual nerve of the cat at various
           constant temperatures and possible time
           course of hypothetical receptor poten-
           tial. Solid lines: threshold depolari-
           zation for triggering of spikes. Dashed
           lines: hyperpolarizing shift in mem-
           brane potential. Further explanation
           see text. From Braun, Bade and Hensel
           (1979).

Fig. 13 shows some hypothetical receptor mechanisms (Bade et al.,
1978; Braun, Bade and Hensel, 1979). The resting potential depends
on the activity of an electrogenic $Na^+ K^+$ pump, the oscillations on
periodic changes of the Na^+/K^+ permeability ratio. The sodium-potas-
sium permeability ratio is assumed to be additionally affected by a
voltage- and temperature-dependent effect of calcium on the potassium
permeability. It is further assumed that a nonlinearity of the calci-

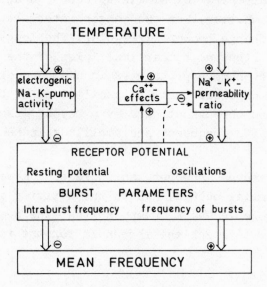

Fig. 13. Hypothetical mechanisms of temperature
transduction of bursting cold receptors.
Further explanation see text. From Braun,
Bade and Hensel (1979).

um-potassium system related to the membrane potential exists, and
that this may act as a negative feedback system: depolarization acti-
vates the calcium-potassium system which, in turn, causes repolariza-
tion.

EFFECT OF CALCIUM AND EDTA

It has long been known that i.v. injection of calcium causes a warm
sensation (for references see Hensel, 1953). Having found that calcium inhibits the activity of cold receptors and enhances that of warm
receptors (Hensel and Schäfer, 1974), we tried to study in detail the
effects of increased and decreased calcium levels on the static and
dynamic behavior of cold receptors. The cat's nose seemed particularly suitable since this receptive field contains bursting as well as
nonbursting cold fibers.

I.v. infusion of calcium (ca. 0.3 mg Ca^{++} min^{-1} kg^{-1}) led to a considerable decrease in the static discharge frequency of cold fibers
and a shift of the maximum to lower temperatures (Schäfer, Braun and
Hensel, 1978). Lowering the calcium level by i.v. infusion of the

chelating agent EDTA (bisodium salt of ethylene diamine tetra acetic
acid, ca. 3 mg min^{-1} kg^{-1}) caused a marked increase in the static
frequency, the effect being greater at higher temperatures.

When cold fibers without static bursting are rapidly cooled, a transient burst discharge during the decreasing phase of the dynamic
overshoot is seen in most cases, the sequence of bursts having the
form of a damped oscillation (Braun and others, 1978). After admin-
istration of Ca^{++}, the dynamic bursts are completely abolished
(Schäfer, Braun and Hensel, 1978), whereas the nonbursting initial
overshoot remains unchanged (Fig. 14).

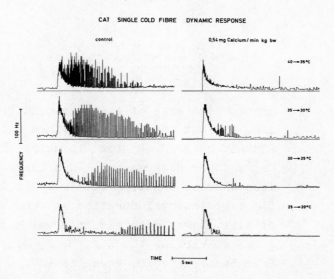

Fig. 14.   Instantaneous frequency of a single cold
            fiber from the cat nose during rapid
            cooling of the skin. Left diagrams: con-
            trols. Right diagrams: i.v. infusion of
            Ca^{++}. From Schäfer (1979, unpublished).

It has been found that calcium also repeals static bursting of lingual cold fibers in cats, while EDTA increases the number of spikes
per burst (Pierau, Ullrich and Wurster, 1977). A detailed analysis
of the discharge of bursting cold fibers in the cat's nose shows
(Fig. 15) that with 2.88 mg Ca^{++} min^{-1} kg^{-1}, the burst discharge is
transformed into a regular sequence of impulses, the intervals of
which correspond to the burst periods, not to the much shorter intra-

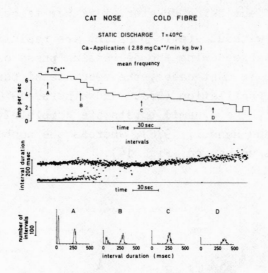

Fig. 15.   Static discharge of a single cold fiber
            from the cat nose at 40°C before (A)
            and after (B, C, D) i.v. infusion of
            Ca^{++}. Upper diagram: mean discharge
            frequency as function of time. Middle
            diagram: interval duration as function
            of time. Lower diagram: interval histo-
            grams at various times A, B, C, D.
            From Schäfer (1979, unpublished).

burst intervals. When the calcium level is lowered by i.v. infusion
of EDTA (3.1 mg min^{-1} kg^{-1}), the opposite occurs: a nonbursting cold
fiber discharge can be transformed into a bursting one (Schäfer et
al., 1979), the intervals of the regular discharge again corresponding to the burst period (Fig. 16).

The effect of EDTA (0.61 mg min^{-1} kg^{-1}) on another cold fiber at con
stant temperatures of 40°C to 15°C is shown in Fig. 17. At higher
temperatures, the discharge frequency increases, and the sequence of
impulses becomes more regular, as seen in the narrower interval histogram. At 35°C, there is practically no change, while at lower temperatures typical bursting occurs after EDTA. In this case, the intervals of the burst periods are considerably longer than the intervals of the regular discharge.

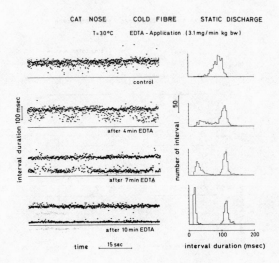

CAT NOSE   COLD FIBRE   STATIC DISCHARGE

T=30°C   EDTA - Application (3.1 mg/min kg bw)

Fig. 16.   Static discharge of a single cold fiber
from the cat nose at 30°C. Left: interval duration as function of time. Right:
interval histograms. Uppermost diagrams:
controls. Following diagrams: 4, 7 and
10 min after i.v. infusion of EDTA.
From Schäfer (1979, unpublished).

If the burst parameters, such as frequency of bursts, impulses per
burst, and intraburst frequency are plotted as a function of constant
temperature, there is no essential difference between naturally bursting cold fibers in the cat's tongue and cold fibers in the cat's nose
with artificial induction of bursting by EDTA (Schäfer and others,
1979). This suggests that the underlying processes are the same. However, the significance of our results for the transduction mechanisms
of cold receptors, in particular the generation of oscillations, must
be left to future experimentation and speculation.

CONCLUDING REMARKS

May I close my presentation with some personal remarks. During my
investigations of intracutaneous temperature and thermal sensations
(Hensel, 1949), it became highly desirable to combine this work with
electrophysiological studies of thermoreceptors. I knew that the only
scientist who had recorded action potentials from specific thermosen-

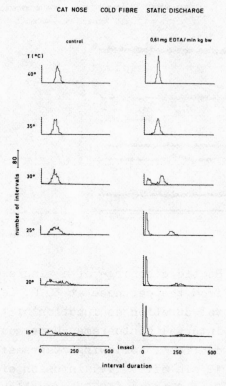

Fig. 17.   Interval histograms from a single cold
           fiber in the cat nose before (left) and
           after (right) i.v. infusion of EDTA at
           various constant temperatures. From
           Schäfer (1979, unpublished).

sitive fibers was Yngve Zotterman (1935, 1936). Exactly 30 years ago
I was fortunate enough to meet him at the congress of the German
Physiological Society in Göttingen in August 1949, where he gave a
paper on "Der Wassergeschmack des Frosches" (Zotterman, 1950). At
this meeting Yngve Zotterman invited me to work in his laboratory.

Before leaving for this symposium, I looked again at the old papers
Zotterman and I published at that time (Hensel and Zotterman, 1951).
I was quite surprised to find that many of the present questions are
found already 30 years ago: the steady discharge, the across-fiber
pattern, the burst discharge, the correlation between temperature
sensation and neural events, the central processing of thermal information, the transduction mechanisms of thermoreceptors - to mention

but a few problems. Some of these are still unsolved.

If any progress in the field of thermoreception has been made since
then, it was not possible without Zotterman's pioneer work. At this
occasion, I would like to thank my friend and colleague Yngve
Zotterman for all I have learned from him and for the inspiration
and encouragement he has always given to me.

REFERENCES

Bade, H., H. A. Braun, H. Hensel, and K. Schäfer (1978). J. Physiol.
    (London), 284, 83 P.

Barker, J. L., and H. Gainer (1975). Brain Res., 84, 461 - 477 and
    479 - 500.

Beste, R. (1977). Perzeption statischer thermischer Reize beim Men-
    schen. Inaug.-Diss. Marburg.

Beste, R., and H. Hensel (1977). Pflügers Arch., 368, R 47.

Braun, H. A., H. Bade, and H. Hensel (1979). Pflügers Arch., in press.

Braun, H. A., H. Bade, K. Schäfer, and H. Hensel (1978). Pflügers
    Arch., 373, R 67.

Dostrovsky, J. O., and R. F. Hellon (1978). J. Physiol. (London),
    277, 29 - 47.

Duclaux, R., K. Schäfer, and H. Hensel (1979). J. Neurophysiol., in
    press.

Dykes, R. W. (1975). Brain Res., 98, 485 - 500.

Eckert, R., and H. D. Lux (1976). J. Physiol. (London), 254, 129 -
    151.

Erickson, R. P., and D. A. Poulos (1973). Brain Res., 61, 107 - 112.

Gola, M. (1976). Experientia, 32, 585 - 588.

Hensel, H. (1949). Ber. ges. Physiol., 135, 468 - 469.

Hensel, H. (1973). Cutaneous thermoreceptors. In: Handbook of sensory
    physiology, vol. 2, Somatosensory system, ed. A. Iggo. Springer,
    Berlin, Heidelberg, New York, pp. 79 - 110.

Hensel, H. (1974). Encyclopaedia Britannica, 328 - 332.

Hensel, H. (1976). Correlations of neural activity and thermal sensa-
    tion in man. In: Sensory functions of the skin, ed. Y. Zotterman.
    Pergamon Press, Oxford, New York, pp. 331 - 353.

Hensel, H. (1977). INSERM, 75, 39 - 56.

Hensel, H., and K. K. A. Boman (1960). J. Neurophysiol., 23, 564 -
    578.

Hensel, H., and A. Iggo (1971). Pflügers Arch., 329, 1 - 8.

Hensel, H., and K. Schäfer (1974). Pflügers Arch., 352, 87 - 90.

Hensel, H., and K. Schäfer (1979). Pflügers Arch., 379, R 56.

Hensel, H., and Y. Zotterman (1951). Acta physiol. scand., 23, 291 -
    319.

Iggo, A. (1969). J. Physiol. (London), 200, 403 - 430.

Iggo, A. (1970). The mechanisms of biological temperature reception.
    In: Physiological and behavioral temperature regulation, ed. J. D.
    Hardy, A. P. Gagge and J. A. J. Stolwijk. Charles C. Thomas,
    Springfield, Ill., pp. 391 - 407.

Iggo, A., and R. L. Ramsey (1976). Thermosensory mechanism in the
    spinal cord of monkeys. In: Sensory functions in the skin of pri-
    mates, ed. Y. Zotterman. Pergamon Press, Oxford, New York, pp. 285
    - 385.

Johnston, D. (1976). Brain Res., 107, 418 - 424.

Konietzny, F., and H. Hensel (1975). Pflügers Arch., 359, 265 - 267.

Konietzny, F., and H. Hensel (1977). Pflügers Arch., 370, 111 - 114.

Meech, R. W., and N. B. Standen (1975). J. Physiol. (London), 249,
    212 - 239.

Pierau, F. - K., P. Torrey, and D. Carpenter (1975). Pflügers Arch.,
    359, 349 - 356.

Pierau, F. - K., J. Ullrich, and R. D. Wurster (1977). Proc. int.
    Union physiol. Sci., 13, 597.

Poulos, D. A. (1975). Central processing of peripheral temperature
    information. In: The somatosensory system, ed. H. H. Kornhuber.
    Thieme Verlag, Stuttgart, pp. 78 - 93.

Sand, A. (1938). Proc. roy. Soc. B., 125, 524 - 553.

Schäfer, K., H. A. Braun, H. Bade, and H. Hensel (1979). Pflügers
    Arch., 379, R 40.

Schäfer, K., H. A. Braun, and H. Hensel (1978). Pflügers Arch., 373,
    R 68.

Schmidt, W. (1976). Die Verarbeitung thermorezeptiver afferenter Sig-
    nale im Trigeminuskern der Katze. Inaug.-Diss. Marburg.

Zotterman, Y. (1935). Skand. Arch. Physiol., 72, 73 - 77.

Zotterman, Y. (1936). Skand. Arch. Physiol., 75, 105 - 119.

Zotterman, Y. (1950). Ber. ges. Physiol., 139, 182 - 183.

ACKNOWLEDGEMENT

This work was supported by the Deutsche Forschungsgemeinschaft.

TEMPERATURE SENSITIVITY IN HUMAN AND NON-HUMAN PRIMATES

D. R. Kenshalo, Sr., J. Greenspan, A. Rózsa and H. H. Molinari

The Psychobiology Research Center and The Department of Psychology,
Florida State University, Tallahassee, Florida, USA

ABSTRACT

Comparisons are made between measurements of the detectability of warm and cool
stimuli, applied to the palm of humans and monkeys, as functions of the temperature
to which the skin of their palms was adapted. Comparisons between warm and cool
stimuli and the area of the stimulator surface are also made. The similarity of
the functions across species suggests that the monkey may serve as an adequate
model for the human thermosensing system.

KEYWORDS

Signal detection; thermal sensitivity of humans and monkeys; thermal sensitivity
and adapting temperature; thermal sensitivity and spatial summation.

One of the most interesting and challenging topics in the study of sensory processes is the code used by the nervous system to transduce and conduct information
about environmental events to the central nervous system. Possible candidate
neural codes (Perkel and Bullock, 1969) in subhuman species are identified by cor-
relating specific aspects of behavior and the concomitant neural events. However,
when candidate neural codes for sensation are sought, it becomes necessary for
human subjects to define the sensation. Notable attempts in this direction have
been pioneered by Hensel and Boman (1960), Borg and colleagues (1967) and, more
recently, the microneurography methods developed by Hagbarth and Vallbo (1967).
[For a recent review of these results see Kenshalo (1979).]

However, with rare exception, these attempts have been confined to correlations
between sensation and neural activity in primary afferent fibers. With these
methods there is little opportunity to observe the signal processing that may occur
at higher centers along the sensory pathway.

An alternative to this direct, but somewhat limited, approach is to establish a
subhuman model for the sensory system under investigation. It then becomes possible to measure, with psychophysical methods, variations in human sensation produced by changes in stimulus conditions; and to correlate these with variations in
neural activity, at several synaptic levels, in the subhuman model with the variations produced by parallel changes in stimulus conditions.

Such correlations rest on the assumption that human sensation is produced by the

87

same set of neural events that can be observed in the subhuman model. Greater confidence may be placed in such correlations if behavioral measurements of the sensory capacity of the subhuman model demonstrate a close correspondence to those of
humans; when nearly identical conditions of stimulation and measurement are used.

To demonstrate this confidence, this study describes measurements of thermal sensitivity of humans and rhesus monkeys. The measurements were obtained by the same
psychophysical method (signal detection); with similar rewards (1 cent or 0.64 ml
of apple juice); and identical punishments (15 sec time-out); for the same stimulus
intensities; adapting temperatures; and stimulator area. We believe that the
signal detection technique is the method of choice because it yields a relatively
pure index of the sensory events, free of motivational and other response biases.

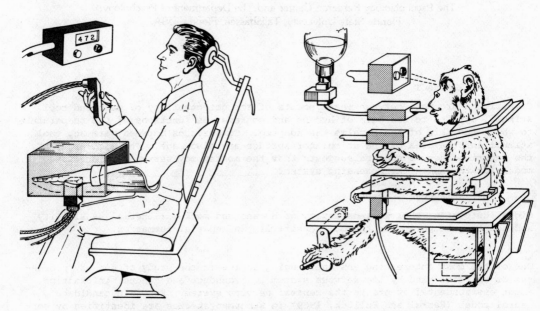

Fig. 1. The experimental arrangements for measurements of
detectability of warm and cool stimuli in humans and monkeys.
Individual trials were initiated by the subject pushing a
button or pulling a lever. Correct identification of the
presence of a change of temperature was rewarded by a penny
(human) or a squirt of apple juice (monkey). Incorrect
identification resulted in a 15 sec time-out.

The experimental arrangements are shown in Fig. 1. The subjects participated in
daily experimental sessions of 1 to 1.5 hrs duration using the yes-no signal detection paradigm described by Molinari, Rózsa, and Kenshalo (1976). The behavior
of both humans and monkeys was shaped so that each subject pushed a button (human);
or pulled a lever (monkey) to initiate a trial; held it for a minimum 3 sec; then
released it if he detected a change in the the temperature of the 7.2 cm^2 thermal
stimulator (Kenshalo and Bergen, 1975) in contact with his left palm. This response was scored as a <u>hit</u>. If no temperature change occurred the subject held
the button (lever) for an additional 3 sec before releasing it. This was scored
as a <u>correct rejection.</u> Either response resulted in a reward. If, on the other
hand, the subject released the button (lever) during the second 3 sec period, even
though no temperature change occurred it was scored as a <u>false</u> <u>alarm</u>. Or, if a

temperature change occurred but the subject failed to release button (lever), a miss was scored and the subject was given a 15 sec time-out. "Trial in progress" was indicated to the subject by a green light and "time-out" was indicated by a white light. During "time-outs" the button (lever) was inoperative.

Temperature changes were delivered at a rate of 1°C/sec from adapted skin temperatures of 28, 30, 33, 36, 38, and 40°C. The intensities of cool stimuli were -0.08, -0.1, -0.2, -0.3, and -0.5°C except at adapting temperatures of 36, 38, and 40°C where the intensity series did not start below -0.2°C. The intensities of warming stimuli were 0.2, 0.25, 0.3, 0.5, and 0.8°C except at the 36, 38, and 40°C adapting temperatures where a sixth intensity of 0.15°C was added.

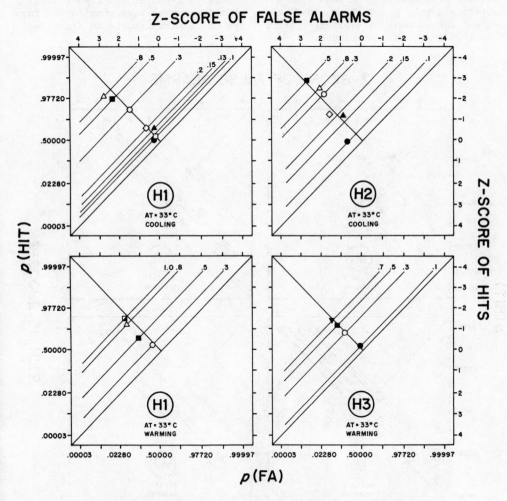

Fig. 2. Human receiver operator characteristic (ROC) curves for several intensities (shown along the top of each graph) of cooling (upper) and warming (lower) from a 33°C adapting temperature. The measure of sensitivity d'e, is measured in Z score units along the minor diagonal. The larger the d'e, the greater the sensitivity of the subject to the intensity of the temperature change. Similar ROC curves for each subject were obtained at the remaining adapting temperatures.

Dan R. Kenshalo, Sr. *et al.*

Proportions of hits and false alarms were computed over 1000 trials for humans and 600 trials for monkeys at each condition of stimulation. While different adapting temperatures were used in a random order, all measurements of cool detection were completed before those for warm detection were obtained.

The same behavioral shaping procedures were used on both humans and monkeys.

ADAPTING TEMPERATURE AND DETECTABILITY

Figures 2 and 3 show the receiver operating characteristic (ROC) curves of detectability in human and monkey subjects for cool and warm stimuli of different intensities of temperature change from a 33°C adapting temperature. These are plotted on double probability coordinates. Similar curves were constructed for cool and warm stimuli of different intensities of temperature change from the other five adapting temperatures.

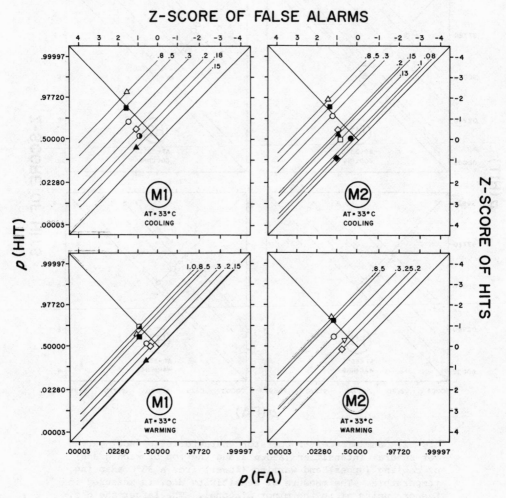

Fig. 3. Same as Fig. 2 except these are detectability curves for monkey subjects.

Estimates of the detectability index, d'e, were determined at each intensity of temperature change for each ROC curve (Pastore and Scheirer, 1974). These values, when plotted as functions of intensity of temperature change, yielded the curves shown in Figs. 4 and 5. The parameter is adapting temperature.

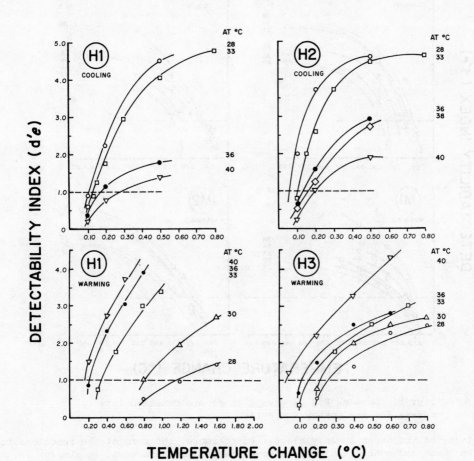

Fig. 4. Plots of the detectability index, d'e, obtained on human subjects as a function of the intensity of the stimulus. The parameter is the adapting temperature shown at the right side of each graph. The interpolated temperature change required to produce a d'e of 1.0 occurs at the point where the d'e curves cross the broken line.

Isodetectability curves, as functions of adapting temperature, were constructed by interpolating the intensity of the temperature change necessary to yield a d'e of 1.0 for each adapting temperature. These are shown in Fig. 6. A d'e of 1.0 results in about a 70 percent correct rate of responding (Egan, 1975) as compared to the 50 percent rate of correct responding employed in classical psychophysical measurement of absolute threshold.

As seen in Fig. 6, detectability of cool stimuli by both humans and monkeys appears to be about equal. One human may have been slightly more sensitive than the other subjects; but the advantage, if any, is small. Both monkeys equaled the

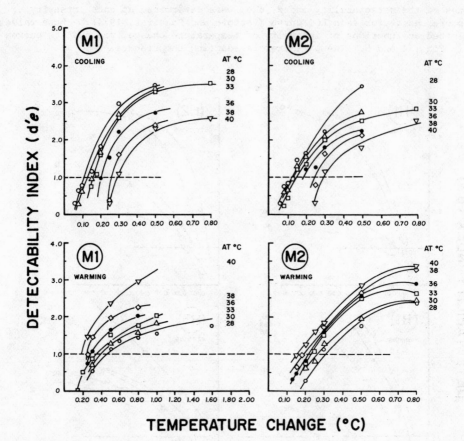

Fig. 5. Same as Fig. 4 except these are detectability
curves for the monkey subjects.

sensitivity of the other human subject. Furthermore, the form of the functions for
both the human and monkey subjects, across adapting temperatures, is similar in
shape.

We conclude that the neural mechanisms by which humans and monkeys perceive cool
stimuli are quite similar under these conditions of measurement. This conclusion
is based on the assumption, as are the remaining conclusions of this presentation,
that similar results from the application of identical procedures imply similar
neural mechanisms.

Detectability of warm stimuli appears to be greater in one human than in either of
the two monkey subjects. Detectability in the other human subject, while similar
to the other subjects at the 33, 36, and 40°C adapting temperatures, was markedly
inferior at the 28 and 30°C adapting temperatures. We believe this difference is
attributable to a difference in the criterion used by this subject as compared to
the other subjects. When the skin of the palm is adapted to low temperatures, a
cool sensation persists no matter how long that temperature is maintained. Episodic
increases in the stimulator temperature are first felt as a decrease in the per-
sisting cool sensation. This has been defined as the "change threshold" (Kenshalo,
1970). As the intensity of each warming episode increases, the subject finally

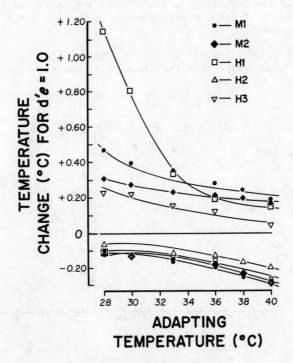

Fig. 6. Isodetectability curves (d'e=1.0) for the
three human and two monkey subjects as functions
of the adapting temperature used.

reports a sensation identical in quality to that experienced with less intense
episodes at higher adapting temperatures and without the "less cool" precursor.
Responses of the one subject, specifically, to the warm quality of the warming episode and the other three subjects to the change threshold, can account for the observed differences in detectability of warm stimuli at the two low adapting temperatures.

Since any apparent difference between humans and monkeys is smaller or identical to
the differences between individuals of the same species, we conclude that the thermal sensing system of monkeys and that of humans function in very similar ways; at
least as far as the affect of adapting temperature upon the detectability of brief
warm and cool stimuli is concerned.

SPATIAL SUMMATION AND DETECTABILITY

The area of application of the temperature change is also a major variable that influences the thermal sensitivity of humans. Studies that have used primarily
classical psychophysical methods applied to human subjects have shown that warm
stimuli summate almost perfectly over area at, or near, threshold intensities.
That is, the threshold intensity decreases by half when the area of application is
doubled on the forehead (Hardy and Oppel, 1937; Kenshalo, Decker, and Hamilton,
1967; Marks and Stevens, 1973) or forearm (Kenshalo and others, 1967). Spatial
summation for warm stimuli of near threshold intensity can be expressed by the
equation:

$$A = kI^{-b} + c$$

where A and I are area and intensity, respectively, k is a scaling constant, c is the warm threshold for very large areas (300 to 400 cm^2), and b is the exponent that varies between 0.79 and 1.0 depending on the subject, and the methods of stimulation and measurement.

Spatial summation of cool stimuli at, or near threshold, can be described by the same trading function except that the exponent is positive and is approximately 0.5 (Berg, 1978).

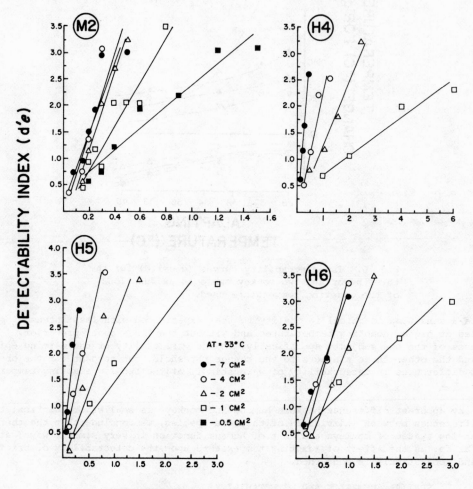

Fig. 7. Plots of the detectability index, d'e, as functions of the intensity of the temperature increases from an adapting temperature of 33°C presented to one monkey (upper left) and three human subjects. The parameter is the area of the palm covered by the thermode. These data were derived from ROC curves similar to those shown in Fig. 2, except area was varied. (These ROC curves are not shown in this paper).

To our knowledge, no one has yet described the spatial summation trading functions of infra-human subjects for either warm or cool stimuli using any of the behavioral methods available. Our aim here is to compare spatial summation on the palm of the humans and monkeys using the same technique described earlier. With the adapting temperature held constant at 33°C, ROC curves were derived from measurements of detectability of warm and cool stimuli applied to the palmar areas of 1, 2, 4, and 7.2 cm² for both humans and monkeys. An additional area of 0.5 cm² was used on the monkey. Estimates of detectability, d'e, were than plotted as functions of the stimulus intensity with area as the parameter. Estimates of detectability for warming from one monkey and three humans are shown in Fig. 7 while those from the same subjects for cool stimuli are shown in Fig. 8.

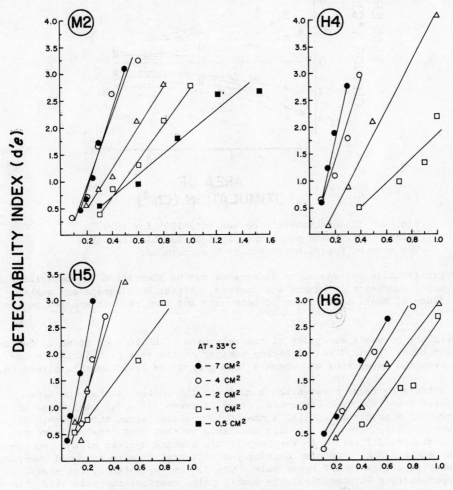

Fig. 8. Same as Fig. 7 except these are functions of the intensity of temperature decreases.

Estimates of the intensities of the stimuli applied to each of the areas required to produce a d'e of 1.0 were obtained from Figs. 7 and 8. These are plotted in Fig. 9.

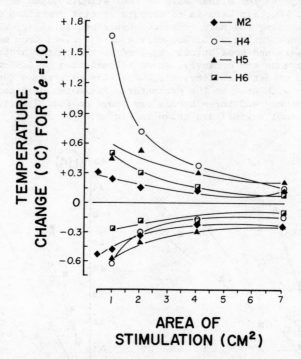

Fig. 9. Isodetectability curves (d'e=1.0) for three
human subjects and one monkey subject as functions of
the area of the palm covered by the thermode.

Two major similarities and one major difference can be identified in comparisons of
thermal spatial summation in humans and monkeys. First, both humans and monkeys
showed evidence of spatial summation of both warm and cool stimuli applied to the
palm.

Second, with the possible exception of the responses of H6 to cool stimuli, detectability functions (Figs. 7 and 8) became steeper as the area of stimulation increased. Even the exception, H6, shows a less pronounced trend in this direction.

The major difference found between the humans and the monkey is that the humans
showed spatial summation over the entire range of areas (1 to 7.2 cm^2) while the
monkey showed no appreciable spatial summation for areas larger than 1 cm^2. It
does not seem likely, based on the data available at this time, that this represents a fundamental difference in the temperature sensing systems of the two species.
It could be explained by a greater density per unit area of warm and cold receptors
in the monkey palm than in the human palm. The 7.2 cm^2 stimulator, for example,
covered approximately 90 percent of the monkey palm, whereas, the same size stimulator covered approximately 10 percent of the human palm. Since we have no receptor counts (except numbers of warm and cold spots on the human palm) for either
human or monkey palm skin, no direct comparison of receptor density is possible.

          ACKNOWLEDGEMENT

This work was supported by NSF Grant BNS76-00820.

REFERENCES

Berg, S. L. (1978). Magnitude estimates of spatial summation for conducted
  stimuli along with thermal fractionation and a case of secondary hyperalgesia.
  Ph.D. Dissertation, Florida State University.
Borg, G., H. Diamant, L. Ström, and Y. Zotterman (1967). The relation between
  neural and perceptual intensity: A comparative study on the neural and psychophysical response to taste stimuli. J. Physiol.,(London), 192, 13-20
Egan, J. P. (1975). Signal Detection Theory and ROC Analysis. Academic Press,
  New York, pp. 81.
Hagbarth, K.-E., and A. B. Vallbo (1967). Mechanoreceptor activity recorded per-
  cutaneously with semi-microelectrodes in human peripheral nerves. Acta physiol.
  scand., 69, 121-122.
Hardy, J.D., and T. W. Oppel (1937). Studies in temperature sensation. III. The
  sensitivity of the body to heat and spatial summation of the end-organ responses.
  J. Clin. Invest., 16, 533-540.
Hensel, H., and K.K.A. Boman (1960). Afferent impulses in cutaneous sensory nerves
  in human subjects. J. Neurophysiol., 23, 564-578.
Kenshalo, D. R. (1970). Psychophysical studies of temperature sensitivity. In W.D.
  Neff (Ed.) Contributions to Sensory Physiology. Academic Press, New York,
  pp. 19-74.
Kenshalo, D. R. (1979). Sensory Functions of the Skin of Humans. Plenum Pub.,
  New York.
Kenshalo, D. R., and D. C. Bergen (1975). A device to measure cutaneous tempera-
  ture sensitivity in humans and subhuman species. J. appl. Physiol., 39, 1038-
  1040.
Kenshalo, D. R., T. Decker, and A. Hamilton (1967). Spatial summation on the fore-
  head, forearm, and back produced by radiant and conducted heat. J. comp. physiol.
  Psychol., 63, 510-515.
Marks, L. E., and J. C. Stevens (1973). Spatial summation of warmth: Influence
  of duration and configuration of the stimulus. Am. J. Psy., 86, 251-267.
Molinari, H. H., A. J. Rózsa, and D. R. Kenshalo (1976). A signal detection
  analysis of rhesus monkey (Macaca mulatta) to cool sensitivity. Perception and
  Psychophysics, 19, 246-251.
Pastoré, R. E., and C. J. Scheirer (1974). Signal detection theory: Considerations
  for general application. Psychol. Bull., 81, 945-958.
Perkel, D. H., and T. H. Bullock (1969). Neural coding. In F. O. Schmitt, T.
  Melnechuk, G. C. Quarton and G. Adelman (Eds.), Neurosciences Research Symposium
  Summaries. MIT Press, Cambridge, Mass., pp. 405-527.

SELECTED REMARKS ABOUT TASTE CELL TRANSDUCTION

L. M. Beidler

Department of Biological Science, Florida State University, Tallahassee,
Florida 32306, USA

ABSTRACT

Knowledge concerning the transduction process of vertebrate taste receptors is very
dependent upon available research techniques. The relative ease of electrophysiological study has resulted in much information concerning the voltage and impedance
changes associated with taste cell excitation. Biochemical studies have been to
date rather disappointing due to the difficulty in obtaining taste cell plasma membrane fractions and to the lack of high specificity of binding. Research on chemoreception of insects and bacteria suggest new ideas for consideration.

KEYWORDS

Transduction; chemoreception; taste; chemical taxis.

FOREWORD

The study of transduction requires quantitative and objective information concerning the response of the taste cells. In 1925, Professor Zotterman was the first to
record from a single nerve fiber and he later used electrophysiological techniques
to investigate numerous sensory systems including taste (Zotterman, 1969). Although
he contributed greatly in many areas of neurophysiology, his greatest and most prolonged interest appears to be in taste. Everyone is familiar with his child-like
curiosity, his spontaneous discussions, and his compassion for others which enliven
any group with which he is associated. As a result, Prof. Zotterman has contributed
greatly not only by his own research, but also by his discussion with others. In
summary, he is both a great scientist and friend.

INTRODUCTION

Chemical stimuli applied to the taste bud may result in the generation of nerve
impulses in the taste nerves associated with the taste bud. The total process that
relates the magnitude of the nerve response to the nature and concentration of the
chemical stimulus is referred to as transduction. It may be quite complex and our
understanding of it is as yet rudimentary. It is the purpose of this paper to
review the highlights of our knowledge concerning this process of transduction.

Our concepts of taste cell stimulation and transduction are strongly influenced by knowledge of such events that occur in other sensory cells. However, even here we have little information. One generality that may or may not be true for taste cells is that the stimulus initiates processes that result in a change of membrane permeability that is associated with the generation of voltage changes across the membrane which in turn are related to the generation of the action potentials of the associated nerves. Since excellent electronic techniques are avaliable to measure such changes, much attention is given to voltage and impedance changes. Unfortunately, techniques for rapid and precise measurement of other physical or chemical changes of the taste cell are not avaliable and such possible changes are either neglected or supposed. Thus, our knowledge of transduction is very much dependent upon both measuring techniques avaliable and the state of knowledge concerning the physical and chemical properties of the taste cell.

STIMULUS TRANSPORT

Stimulus transport has not received much attention by most researchers. The stimulus is usually an aqueous solution flowed over the tongue. A solution of physicochemical hydrodynamics is rather complex but the case of liquids flowing over a flat surface has been described (Levich, 1962). During uniform flow, all the concentration changes occur in a 10-50μ diffusion boundary layer next to the tongue surface even though the velocity of the stimulus stream may increase at larger distances above the tongue surface. The depth of the diffusion boundary layer, δ, can be approximated as:

$$\delta \sim D^{\frac{1}{3}} \, \nu^{\frac{1}{6}} \, (\frac{X}{V_o})^{\frac{1}{2}}$$

where:
D= diffusion coefficient of molecules in water
ν= kinematic viscosity of water (about 10^{-2}cm^{-2}/ sec)
X= distance from leading edge of body
V_o= free stream velocity

Note that the depth of the diffusion boundary layer is a function of both viscosity and diffusion coefficient. If large molecules , such as the proteins that elicit sweetness, are considered as rigid spherical molecules, then Stoke's law can be applied and the diffusion coefficients are found to be inversely proportional to the cube root of the molecular weight. Thus, large molecules must diffuse a smaller distance but at a much slower rate to the taste cells and the latency of response is increased (See Appendix).

The depth of the diffusion boundary at a given point along the tongue surface is noted to also decrease with the square root of the velocity of stimulus flow.

The depth of the unstirred layer above the microvilli may be less than that calculated from the above expression due to the fact that the tongue surface is not smooth and the microvilli are found in taste pores. In fact, the unsmooth surface structure of the tongue implies that calculations of the depth of the unstirred layer must result in very rough approximations.

Electrophysiological experiments of flow rate have been confounded by the fact that the taste buds are distributed over a relatively large area of the tongue surface. As the velocity of stimulus flow increases, the number of taste cells activated per unit time at the beginning of the flow also increases so that the initial summated response is greater. This is emphasized by the rapid initial decline (half life of about 800 msec) of single nerve fiber activity in response to a constant NaCl stimulus (Beidler, 1953). For example, consider three identical taste buds separated by 1 mm each along the direction of the stimulus flow path and assume a linear flow velocity of 0.5 mm/sec. The response can then be diagrammed as:

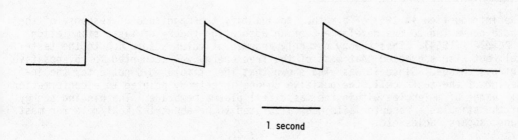

1 second

If the linear flow is increased to 5mm/sec, then the initial summated response is almost three times larger or:

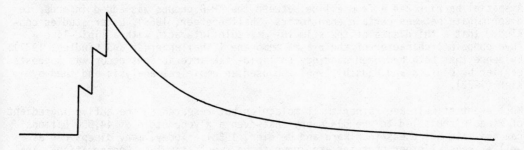

If an unstirred layer is precluded by the roughness of the tongue surface, then only the geometry of the taste pore need be considered for diffusion. The depth of the pore is usually 3-5 μ and the microvilli extend to its outer surface. If a conservative estimate of 10 μ is considered, then calculations reveal that the average concentration within the pore reaches 90% of the applied NaCl concentration within 30 msec. The tip of the microvilli would be stimulated in a much shorter time. This is in contrast to the 18-50 msec electrophysiological latency. If the large sweet protein, miraculin, is considered as a rigid sphere, then calculations indicate it would take over 25 sec to reach 90% of the applied concentration at a depth of 10μ!

It is interesting that some diseases, such as those associated with Zinc deficiency, produce hyperkeratosis of the tongue surface. The depth of the taste pore, and thus the diffusion path, may be as high as 100μ or more (Ewen and Beidler,1979). Thus, it would take NaCl over 7 sec to reach its applied concentration at the bottom of the pore where the microvilli are found. This condition would result in very long response latencies, slow rise of response magnitude, and higher taste thresholds.

It should be noted that single fiber responses to small molecules or ions such as

Na^+ or H^+ start with maximum frequency and then decline. No build-up frequency is noted after the initial response is initiated.

INITIAL BINDING ASSOCIATED WITH RECEPTOR STIMULATION

Structure - activity relationships

The introduction in 1951 of a method to quantify the magnitude of response of the taste nerve led to the development of an adsorption theory of taste stimulation (Beidler ,1954). This theory not only assumed stimulus adsorption to the taste cell but also suggested that most of the free energy was accounted for by positive entropy change. Since it was also shown that the stimulus did not enter the in- terior of the taste cell, the positive change in entropy pointed to a conformation- al change of molecules within the taste cell plasma membrane. The binding energy of the stimulus - receptor site was calculated to be about 1-2 kcal/mole for most ions, sugars, acids etc.

Shallenberger and Acree (1967) considered the weak forces to be a result of hydro- gen bondong in the case of substances sweet to man. They studied the structures of a large number of sweet substances and concluded that a necessary requirement is two sites within the stimulus molecule that are separated by about 3.0 Å and which can hydrogen bond to some complementary structure within the plasma membrane. A spatial barrier 3-4 Å from a line between the AH-B groups was added in order to discriminate between certain enantiomers (Shallenberger, 1969). Later studies con- cluded that a third site of the stimulus molecule interacts with a lipid-like (hydrophobic) character of the plasma membrane (Shallenberger and Lindley, 1977). Evidence that both hydrogen bonding and lipid-like interactions occur was suggested earlier by Deutsch and Hansch (1966) who used an empirical analysis and then by Kier (1972).

Most sweet stimuli are rather small molecules but research on the active ingredient of miracle fruit led to the discovery that even a glycoprotein of 44,000 Daltons can initiate sweetness (Kurihara and Beidler, 1968). Today, many dipeptides as well as several other proteins are known to be sweet stimuli. Conformational anal- ysis of the unusual dipeptide, methyl ester aspartic acid phenyl alanine, suggest- ed the following receptor site (Temussi, Lelj, and Tancredi, 1978).

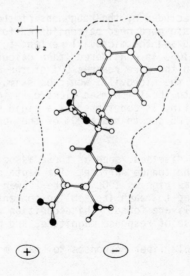

A number of different sweet stimuli appear to fit properly in the above receptor
site. However, caution must be expressed since the most probable conformation of
the dipeptide was used to determine the geometry of the receptor site whereas the
less probable conformations may actually be those used in taste bindings. It is
interesting to note that the authors of this analysis not only are concerned with
those molecular attributes that distinguish sweeteners from nonsweeteners, but
also the attribute that determines the intensity of sweetness.

Stimulus molecules can bind to many places within the taste cell plasma membrane
but only interaction at some specific sites initiate a taste reaction. This is
obvious with the hydrogen ion that can react with many carboxyl groups but sour-
ness only occurs when H^+ reacts with carboxyl groups of relatively low pk.
Similarly, it is thought that inorganic ions can react to many different types of
receptor sites that are involved in the generation of saltiness. The narrowness
(3 log units) of the sigmoid curve of a R vs log C plot suggests that most of
the receptor sites can attract the stimulus ion or molecule with about the same
attractive force. Notable exceptions do occur as seen with amino acid binding by
taste cells of catfish (Caprio, 1975). The broad response-concentration curve may
be a result of different receptor sites of varying attractive forces or association
constants (K).

The adsorption equation can be successfully applied to many different species,
from man to insects. It is often confused with the Michaelis-Menton equation used
for enzyme kinetics. The equations are analogs but describe completely different
events and should not be confused. The one describes a process at equilibrium
whereas the other describes a kinetic process.

Vertebrate taste receptor molecules

Structure-function analysis has led to preliminary concepts of the conformation of
the taste cell receptor sites as related above. Is it possible to isolate those
membrane proteins or lipids that interact with the taste stimuli? Many researchers
have tried but no convincing evidence exists that such a specific protein from
vertebrate taste cells has been isolated and characterized. The small quantity of
plasma membrane in a taste cell and the weak binding forces makes the task of
isolation and characterization very difficult. It is well known that many non-
taste cells also adsorb ions or molecules that are considered taste stimuli.
Therefore, researchers have tried to correlate some measure of magnitude of taste
response to the ability of their receptor molecule preparation to bind a given
taste stimulus. What measure of response is most appropriate? Certainly not preference thresholds nor any data from species different from that used in the biochemical preparations.

Much electrophysiological evidence indicates that the number of receptor sites or
(and) the ability to initiate the transduction mechanism varies from one taste
stimulus to another, even those perceived as having similar taste qualities. This
is reflected in both the magnitudes of maximum response, R_S and association constant, K. Thus, the strength of binding of taste stimuli to receptor sites, as
determined biochemically may reveal little insight into taste transduction until
the number and intrinsic activity of the sites are known. Furthermore, thresholds
are dependent upon both the strength of binding ($\sim$K) and the intrinsic effectiveness ($\sim R_S$). For this reason, thresholds would not be expected to necessarily correlate well with biochemically obtained association constants.

The biochemical preparation varies greatly with researchers. Some try to obtain
rather pure preparations of taste buds whereas others simply remove portions of the

circumvallate papilla and compare it to other lingual tissue not containing taste
buds. In many cases the tissue is merely homogenized and no effort is made to
further fractionate and obtain membrane concentrated preparations. Price and
DeSimone (1977) have reviewed the literature on taste receptor proteins and presented an analysis from the viewpoint of a biochemist.

Insect taste receptor molecules

An interesting story can be told for insect receptor molecules. Dethier observed
that the legs of the fly, Phormia, contain an enzyme that splits certain sugars
Dethier,1955). The distribution of the enzyme on the fly corresponds to those
areas rich in taste cells (Hansen, 1969). The enzyme was found to an α -gul-
cosidase. The K_m values of a specific glucosidase isolated from the blowfly agree
with values expected from electrophysiological responses of the blowfly to a
series of sugars (Morita, 1972). Furthermore, a sugar receptive mutant of the
fruitfly, Drosophila melanogaster, was shown to have both reductions in electrophysiological responses and hydrolytic activities to compounds that normally
react with the α-glucosidase (Kikuchi, 1975). One may tentatively conclude that
the α- glucosidase is involved some way with the taste receptor response to certain sugars but the exact mechanism is unknown.

Chemotaxis

Since most species respond to chemicals, one may look at perhaps more simple examples of stimulus-cell interactions. Rabbit peritoneal leukocytes stimulated
with the chemoattractant peptide, met-leu-phe, show an increased degradation of
membrane methylated phospholipids (Hirata, 1979). This occurs by activation of
the enzyme phospholipase A_2 which removes the unsaturated fatty acid from the
membrane phospholipids. This alters the arrangement of the lipids across the
membrane and may change membrane fluidity and charge.

A methyltransferase system may also be important in bacterial chemotaxis. Studies
of E. coli mutant behavior and biochemistry led to the hypothesis that Ca++
permeability may control chemotaxis. The pump controlling the Ca++ concentration
within the cell is not affected by the chemotactic stimulus, but rather the gate
is affected. It is thought that excitation closes the gate and adaptation opens
it by methylating the gate molecule. It should be emphasized that these experiments include observations on the total behavior of the bacteria and not just
the sensory input. However, the importance of methylation in chemotaxis of both
bacteria and leukocytes should not be missed by those studying vertebrate taste
cell transduction.

TASTE RECEPTOR POTENTIALS

Little information is known concerning the transductive events between the time
the chemical stimulus is adsorbed to the taste cells and the appearance of receptor potentials. The origin of these potential changes does not appear to be
located at the taste cell microvilli where stimulus adsorption is assumed to
occur, but may be located toward the base of the taste cells. This situation is
not too dissimilar to that of the rod visual cell where Ca++ is thought to be the
means of communication between the site of photon absorption and the site of
origin of the receptor potential on the rod plasma membrane. The means of communi-

cation in the taste cell is not known but propagated physicochemical changes
in the plasma membrane have been suggested (Beidler and Gross, 1971).

Properties of the taste receptor potentials have been well studied by several
researchers. A summary of all the data will not be presented here. However,
it should be emphasized that a correlation of the magnitudes of receptor
potentials with the frequency of nerve action potentials has been suggested but
a direct relationship has not yet been obtained. Simultaneous measurement of
receptor potentials and membrane impedances confirms earlier suggestions that a
change in membrane permeability is involved in taste cell transduction. Experiments by Sato and Beidler (1975) imply that Na++ ions may be important in
these changes of permeability in the frog.

MODELS OF TASTE CELL TRANSDUCTION

Modern theories of taste cell transduction have utilized electrophysiological data
to correlate with the nature and concentration of the taste stimuli. These models
may be both quantitative and predictive and thus are very useful even though
future data may make it necessary to either alter or supplant them.

The oldest of these models quantitatively applied Langmuir's adsorption theory to
the taste process.(Beidler, 1954). The binding constants associated with the
taste stimulus adsorption to the taste receptor sites implied that the binding
strength was very weak (1-2 kcal/mole). Consideration of the taste process suggested that adsorption leads to changes in membrane permeability and membrane
impedance. Phase boundary potentials were not considered to be important since
they would not necessarily alter the electrical gradient within the plasma membrane that was assumed to be important in the transduction process.

Kurihara and colleagues (1974, 1978) and DeSimone and Price (1976) later emphasized the phase boundary potentials in addition to the diffusion potential associated
with the taste receptor membrane. The Japanese group did not assume that a
Hodgkin type diffusion potential is part of the transduction process. In contrast,
the American group assumed that changes in the surface charge of the plasma membrane essentially controls the passage of the ionic taste stimuli across the
plasma membrane.

Whichever model is considered, it should be remembered that the order of effectiveness of a series of taste stimuli change from one species to another and perhaps
also from one taste cell to another within the same species. This can be accounted for if adsorption is assumed and the amount of water of hydration is
dependent upon the electric field strength of the receptor site.

NEURAL ACTIVITY

The transduction process ends with nerve discharge. Very few studies of the
nature of the irregular neural firing of taste fibers have been conducted. In
many ways it is similar to that observed with Type I cutaneous mechanoreceptors
(Horch,Whitehorn and Burgess, 1974). The nerve branching and multiple innervation
of receptors is similar in the taste and Merkel cells. Although neural discharge
is irregular in both systems, occasionally pairs of action potentials with short
intervals are observed, such as in the response of rat taste fibers to sugar
stimulation. Perhaps additional anatomical studies of multiple taste bud innervation and also physiological studies of the synaptic connections will serve to

illucidate the origin of the irregular firing of taste fibers.

SUMMARY

Much of the taste transduction process is still a mystery, as it is for most other sensory systems. What is needed most is more original and new approaches to its study, particularly biophysical. In addition, rational study and criticism of taste research by scientists in other fields is sorely needed. Knowledge of other sensory processes is also helpful. Perhaps this is why the remarks of Professor Zotterman have been so pertinent in the field of taste.

Appendix

Calculated concentration, c, of NaCl and the protein, miraculin, at a depth of 10 μ and 50 μ in an unstirred layer.

MIRACULIN & NaCl

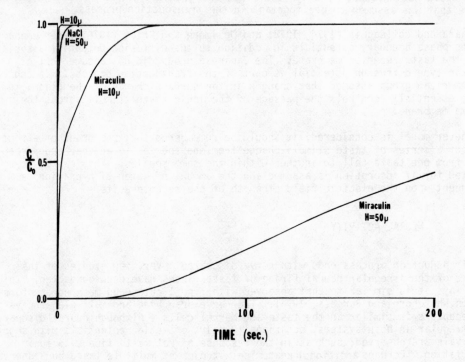

REFERENCES

Beidler, L.M. (1953). Properties of chemoreceptors of tongue of rat.
    J. Neurophysiol. 16, 595-607.
Beidler, L.M. (1954). A theory of taste stimulation. J. Gen. Physiol. 38,
    133-139.
Beidler, L.M., G.W. Gross (1971). The nature of taste receptor sites. In W.D
    Neff (Ed.), Contributions to Sensory Physiology, Vol. 5, Academic Press,
    New York.
Caprio, J. (1975). High sensitivity of catfish taste receptors to amino acids.
    Comp. Biochem. Physiol., 52, 247-251.
DeSimone, J.A., S. Price (1976). A model for the stimulation of taste receptor
    cells by salt. Biophys. J. 16, 869-881.
Dethier, V.G. (1955). Mode of action of sugar-baited fly traps. J. Econ. Ent.
    48, 235-239.
Deutsch, E.W. and C. Hansch (1966). Dependence of relative sweetness on hydro-
    phobic bonding. Nature, 211, 75.
Ewen, J.E., and L.M. Beidler (1979). Hyperkeratosis of the Zinc Deficient Rat
    Tongue Alters Diffusion of Taste Stimuli to Fungiform Taste Buds.
    Submitted to Physiology and Behavior.
Goy, M.F. and M.S. Springer (1978). In search of the linkage between receptor
    and response: The role of a protein methylation reaction in bacterial chemo-
    taxis. In G.L. Hazelbauer (ed), Taxis and Behavior. Chapman and Hall,
    London. pp. 1-34.
Hansen, K. (1959). The mechanism of insect sugar reception, a biochemical in-
    vestigation. In C. Pfaffmann (Ed.), Olfaction and Taste III.
    Rockefeller University Press, New York. pp. 382-391.
Hirata, F., B.A. Corcoran, K. Venkalasobramanian, E. Schiffmann and J. Axelrod
    (1979). Chemoattractants stimulate degradation of methylated phospholipids
    and release of arachidonic acids. Proc. Natl. Acad. Sci. USA 76, 2640-
    2643.
Horch, K.W., D. Whitehorn and P.R. Burgess (1974). Impulse generation in type I
    cutaneous mechanoreceptors. J. Neurophysiol. 37, 267-281.
Kamo, N., M. Miyake, K. Kurihara, and Y. Kobatake (1974). Physicochemical
    studies of taste reception. I. Model membrane stimulating taste receptor po-
    tential in response to salts, acids and distilled water. Biochim. Biophys.
    Acta 367, 1-10.
Kier, L.B. (1972). A molecular theory of sweet taste. J. Pharmaceut. Sci. 61,
    1394-1397.
Kikuchi, T. (1975). Genetic alteration of insect sugar reception. In D.A. Denton
    and J.P. Coghlan (Eds.), Olfaction and Taste V. Academic Press, New York.
    pp. 27-32.
Kurihara, K. and L.M. Beidler (1968). Taste modifying protein from miracle fruit.
    Science. 161, 1241-1243.
Kurihara, K., N. Kamo, and Y. Kobatake (1978). Transduction mechanism in chemo-
    reception. Adv. Biophys. 10, 27-95.
Levich, V.G. (1962). Physicochemical Hydrodynamics. Prentice Hall, New Jersey.
Marita, H. (1972). Properties of the sugar receptor site of the blowfly. In
    D. Schneider (Ed.), Olfaction and Taste IV. Wissenschaftliche
    Verlagsgesellschaft MBH, Stuttgart, pp. 357-363.
Price, S. and J.A. DeSimone (1977). Models of taste receptor cell stimulation.
    Chemical Senses and Flavor 2, 427-456.
Sato, T. and L.M. Beidler (1975). Membrane resistance change of the frog taste
    cells in response to water and NaCl. J. Gen. Physiol. 66, 735-763.
Shallenberger, R.S. and T.E. Acree (1967). Molecular theory of sweet taste.
    Nature 216, 480-482.

Shallenberger, R.S. and M.G. Lindley (1977). A lipophilic-hydrophobic attribute
    and component in the stereochemistry of sweetness. Fd. Chem. 2, 145-153.
Temussi, P.A., F. Lelj, and T. Tancredi (1978). Three-dimensional mapping of the
    sweet taste receptor site. J. Med. Chem. 21: 1154-1158.
Zotterman, Y. (1969). Touch, Tickle and Pain. Part I, Pergamon, London.

THE SENSE OF TASTE AND BEHAVIOR

C. Pfaffmann

The Rockefeller University, 1230 York Avenue, New York,
New York 10021, USA

ABSTRACT

This paper describes how the sense of taste controls appetitive and instrumental
behavior but also that behavioral methods can reveal how animals perceive taste
stimuli. Such methods permit direct comparison of animal psychophysics with sensory electrophysiology in the same organism. The behavioral classification of
qualities according to the basic taste stimuli corresponds very well with that
derived from the electrophysiological best-stimulus fiber designation in rats and
hamsters. Taste stimuli not only elicit preference and aversion behaviors; with
appropriate training schedules they can reinforce learning of a variety of instrumental responses. Neurological experiments are described showing that the oro-
mimetic components of the sweet preference are organized in the ponto-bulbar lower
brainstem and are present in chronic decerebrate mammals. These response patterns,
normally present at birth, can be modified in normals by learning but not in
chronic decerebrates. The sense of taste is uniquely suited for analyzing the role
of sensory stimuli in motivation and learning, in sensory affect and hedonic processes, as well as its sensory physiology.

KEYWORDS

Taste electrophysiology; taste preference; taste aversion; oro-mimetic responses;
chorda tympani; decerebrate; best-stmulus class; hedonic response; conditioned
taste aversion; reinforcement.

INTRODUCTION

It is a great personal pleasure to join in this tribute to Yngve Zotterman. In
addition to his many other interests in sensory physiology, he, of course, made an
early and continuing contribution to the study of the sense of taste. It was his
enthusiastic sponsorship of the first International Symposium on Olfaction and
Taste in Stockholm in 1962 that gave ISOT such a good start that it has carried on
through the VIth Symposium in 1977 with increasing membership and richness of pro-
gram. In the 1960s the field of chemoreception was still relatively a small scientific enterprise. I recall that it was at an informal meeting at my laboratory at
Brown University, with Yngve Zotterman and Lloyd Beidler that we conceived the
ISOT idea. The series as a whole prospered through Yngve's enthusiastic support as

109

a member of the Council and later President of the International Union of Physiological Sciences. I succeeded him as chairman of the Commission on Olfaction and Taste and Lloyd Beidler succeeded me.

Yngve and I share another common bond - the "Cambridge connection," although we did not overlap there. His classic work with Adrian on single sensory fibers occurred in 1925. I arrived to start my taste recording as part of a Ph.D. thesis under Adrian in 1937 and gave my first report at the Oxford meeting of the Physiological Society in 1939. Along the way since then, there have been many happy personal associations with Yngve and Brita and members of their family.

TASTE AND BEHAVIOR

Today I have chosen to talk about some of the behavioral aspects of taste and their physiological bases. As a physiological psychologist it is both the behavior and the physiology underlying that behavior that has intrigued me. Taste, like olfaction, is a sense in which certain stimuli from the time of birth elicit a series of stereotyped but modifiable responses of attraction and preference on the one hand, or revulsion and rejection on the other, most often in relation to food and fluid intake and appetite. In certain aquatic species where taste is distributed more widely over the body surface, it may function as an exteroceptor. In land inhabiting vertebrates, and I shall restrict my discussion mainly to mammalian forms, taste is a special visceral sense on the tongue and in the oral cavity where, nevertheless, external stimuli in the environment can easily stimulate it. Taste shares some properties with exteroceptors, it shares others with the interovisceral receptors, and as we shall see it has a rather special relationship to the visceral world. It also has a special relation to hedonic value, to sensory affect, what I have referred to as "the pleasures of sensation" (1960). Most taste stimuli can be classified as pleasant or unpleasant; they are less often merely neutral. Cabanac (1971) has shown that pleasurable judgments of a sweet solution gradually shift to unpleasantness if the solutions are ingested but do not if merely tasted periodically for an equivalent period but not swallowed.

Of course Stimulus intensity may be a parameter in this characterization. Strong stimuli, as in other modalities may be unpleasant but not necessarily so, and certain weak stimuli like quinine may be unpleasant and rejected. It is this plus-minus hedonic dimension that is readily tapped by the appropriate choice of stimuli or the modification of their molecular configuration. Finally, by virtue of these hedonic qualities, taste stimuli can function as reinforcers of behavior. The non-hungry rat will learn to press the bar of an apparatus that delivers a brief sip of nutritive or non-nutritive sweetener, the hungry rat will press harder. The new born baby will suck more vigorously the stronger the sugar solution (Nowlis and Kessen, 1975) and will differentially learn to turn the head to the right or left depending on which nipple delivers the sweeter fluid (Papousek, 1967).

These relations are strong and so to speak "hard wired" but they are not inflexible. One of the most interesting discoveries of recent years is the conditioned taste aversion by John Garcia and colleagues (1955; 1974). The saccharin that was once so attractive can become repulsive after a single pairing with visceral distress and nausea, and although the similarly dramatic converse, that is, rendering a repulsive stimulus palatable is not so much a one time thing, but a more slowly acquired change from unpalatable to palatable. In this regard I should like to quote from Dr. Judson Brown (1955) who so aptly noted:

> "Straight whiskey, when first ingested, typically effects
> rather violent defense reactions. Because of this, the

novice drinker usually begins with sweet liquers, "pink
ladies," and wines, and slowly works his way through a series
of beverages characterized by the gradual disappearance of
cola and ginger ale additives. Finally, only plain water
or even nothing need be mixed with the raw product. To the
hardened drinker, straight whiskey does not taste bad--"not
bad at all!" (It is thus that a product euphemistically
labelled "neutral spirits" becomes indeed psychologically
neutral).

More recently Paul Rozin (1978) has been examining with still considerable mys-
tification,   the mechanisms of the acquisition of the taste for chile peppers.
His observations indicate that this must be learned in the tropical climes and cul-
tures.   In the beginning children in these regions show strong aversive reactions
and reflexes of crying, sneezing and revulsion at their first encounter with them.
But the chili-hating infant turns into a chili-loving adult.

However, so much for generalities. Let us proceed on a combined physiological-
behavioral review of some aspects of taste. To set the stage let me begin with a
couple of figures from my taste collection.

Fig. 1. A neolithic "sweet tooth."
Cave painting of an individual
robbing a store of wild honey
(Arana Cave, Spain, reproduced
from Deerr, 1949)

Fig. 2.   A Lincoln Park giraffe using its tongue instrumen-
          tally first to obtain and then savour the gingerly
          offered marshmallow (Credit, United Press Inter-
          national Photo, 1973)

These are both examples of the sweet tooth so to say. I shall concentrate on that
but let me remind you that sugar alone is not the only taste attractant. Salt is
another well known, much studied appetite over the years, particularly stimulated
by the early work of Curt P. Richter (1942) and though we know much about it now,
there are still certain unanswered questions as to the interaction and role of
sensory and central mechanisms in this process.

More typical of the laboratory tests of taste preferences and aversions is the
well-known Richter two-bottle preference test. The data typically shows preferen-
tial intake of the solution of interest relative to that of water after a 24 hour
exposure. Preference or aversion thresholds are readily obtained by an increase
in concentration series as shown in the top panel of Fig. 3. Notice that the
aversion response is rather straight forward. At some concentration the animal
takes less of the solution than water and ultimately may take none or very little
of the quinine or acid. The picture for preferred substances is somewhat more
complex. At some threshold value the organism takes more of the tastant than the
water and with increasing concentration takes more and more of it to some peak
preference value. Then, the curve flattens and turns down as the animal begins to
take less and less of it on the so-called aversion limb. But is that a true
aversion?

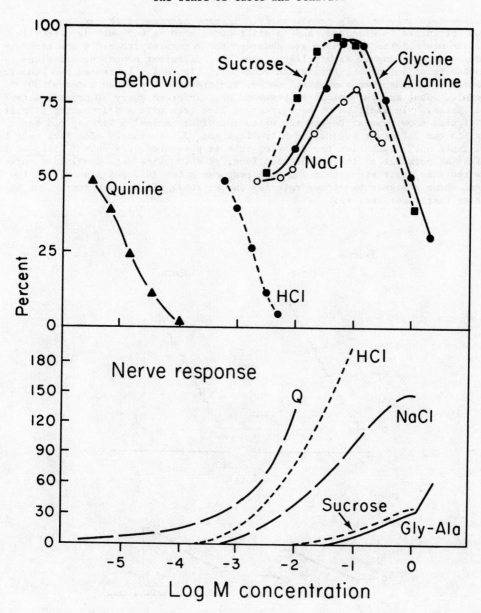

Fig. 3. Behavioral aversions and preferences to taste (upper
        panel) and composite VII and IX nerve discharges
        (lower panel). (From Pfaffmann, 1977).

There are now many variants on the methods for studying taste preferences which
cast doubt that the descending limb does reflect the true aversion in all cases.
Some of the methods are short-term to minimize post-ingestive effects. Some methods
interpose an instrumental response such as a bar press that may deliver a drop of

solution regularly or only occasionally.   In some methods, lick rate and not
volume of intake is measured, and in still other more recent methods, specific
oral and oro-mimetic responses are observed and accurately recorded and timed by
videotaped recordings.   Each tells us something different about the stimulus-
response relations between taste and behavior.   Figure 4 illustrates two behavior
measures for sugar when a squirrel monkey bar presses on a mixed interval 30
schedule.   That is, the animal is rewarded or reinforced every 30 seconds after
a bar press.   In addition we have recorded lick rate with a drinkometer circuit.
The figures show cumulative records of an individual animal's licking and bar
press to two different sugars, .3 M fructose and .3 M sucrose.   The lick rate has
"ceilinged out"; each time the drinking tube is presented, the animal licks at it
as fast as possible at these concentrations.   The two lick rate cumulative curves
show the same stair steps form and same response rates (520 per minute) yet the
animal shows a higher bar press rate for the sucrose.   It works harder, that is
presses faster for sucrose.

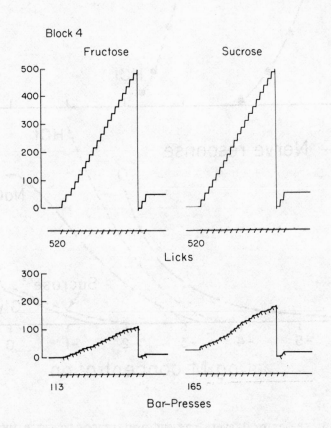

Fig. 4.   Cumulative records of licking and bar-pressing in
           the same animal for two different sugars showing
           unequal instrumental response rates but equal oral
           behavior (lick rates).   Data obtained with the
           assistance of B. McCutcheon and J. Coolsey.
           (Pfaffmann, 1969).

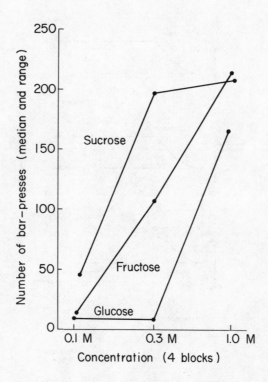

Fig. 5.   Bar pressing rate functions of squirrel monkey for
          three sugars from experiments illustrated in Fig. 4.
          (Pfaffmann, 1969)

The response rate as a function of concentration for three sugars is shown in Fig.5.
Note that these operant response functions do not turn down for the higher sugar
concentrations.   There is no aversion limb to these response curves.   In the two
bottle preference test, there is clearly a turn-down for .3 M and 1.0 M sugars.
Numerous studies have experimentally separated the operation at least of two
factors;   one a go signal arising from the gustatory stimulus, the other a stop
factor arising from the post ingestive influences in the two bottle 24 hour test
as schematized in Figure 6.   Davis and Levine (1977) have, in fact, developed a
mathematical model along these lines and have been able to predict quite accurately
the outcome of short-term 3 minute and long-term drinking behavior.   But other
models of positive-negative determiners of intake functions can be developed.
Sometimes the negative factors may be of mouth origin.   In all such treatments,
taste is the first sensory input in the sequence that ultimately triggers the
behavior.

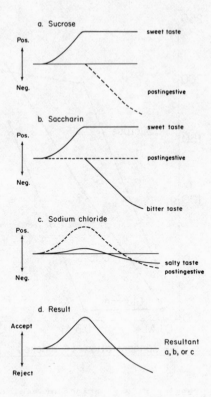

Fig. 6.   Hypothetical interaction of taste and post-ingestive
          factors underlying the typical preference-aversion
          function.   (Pfaffmann, 1969)

But what about the sensory input studied physiologically? The lower curves of Fig. 4
give the electrophysiological functions that can be correlated with behavioral response. The curves are generated from the summated activity in the chorda tympani
nerve as illustrated in Fig. 7. I have purposefully selected this example of the
method of recording which is applicable to both animal and man as Diamant and
Zotterman first showed in 1959. This well known figure shows the reversible blockade of the sweet receptor by gymnemic acid brushed on the tongue. The whole nerve
record is a composite measure of activities in all fibers so that the contribution
from a population of different types of receptors with differing sensitivities can
only be revealed by single fiber analysis.

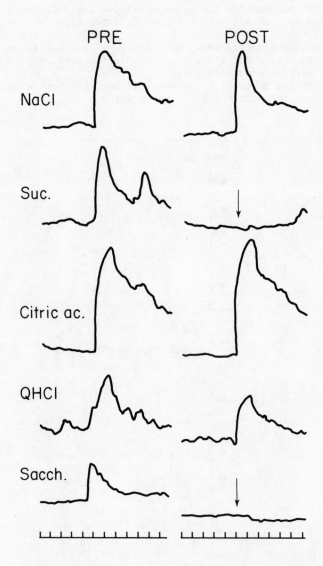

Fig. 7. The summated electrophysiological response of the
human chorda tympani to the indicated stimuli before and after the application of gymnema extract.
Concentrations were 0.2 M NaCl, 0.5 M sucrose,
0.02 M citric acid, 0.002 M quinine hydrochloride,
and 0.004 M sodium saccharin. All responses from
the same patient. (Diamant and colleagues, 1965)

Because of our interest lately in sugar and sweet sensitivity generally, we have
focused our studies on the hamster, a species with a relativelly good sugar sensitivity on its anterior tongue compared to the rat. Like the human, its response to
sugar is subject to reversible blockade by gymnemic acid. I would like to note
here that it was Yngve Zotterman and his colleagues (Anderson and co-workers, 1950)

118   Carl Pfaffmann

who were the first to describe the sugar sensitive taste fibers using the dog and
their suppression by gymnemic acid. Figure 8 illustrates the response profiles for
three sugar-best fibers and a series of sugars, amino acids and other basic taste
stimuli recorded by Dr. Nowlis in my laboratory. The lower figure shows the relative ineffectiveness of sweet stimuli in stimulating acid or salt-best fibers.
Note also the clear difference in response for the D and L forms of phenylalanine
on the sweet sensitive fibers. D is sweet to man, the L form tasteless or bitter,
which seems to parallel the hamster's sensitivity.

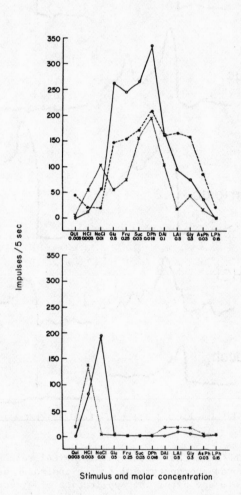

Fig. 8.   The upper panel shows the response profile of 3
          sucrose-best hamster chorda tympani (CT) units to
          basic taste stimuli, other sugars, D- and L-amino
          acids, and, aspartame (AsPh). The concentrations
          indicated were equi-effective on the summated whole
          CT nerve except for AsPh. The lower panel shows
          the response profiles of one acid-best and one acid-
          best fiber to the same stimuli (Pfaffmann, 1978).

But we can do better than simply assume the hamster's taste from the electrophysiology. We can now, in effect, test the animal's sensory world by use of the
conditioned taste aversion procedure. Conditioned taste aversions to a particular
taste are readily and selectively established after one or a few trials which
associate a particular taste quality with the drug induced gastric or visceral distress (Garcia and co-workers, 1974). The specificity of this association provides
a method for testing similarity by the degree of generalization among different
taste substances following conditioning. To run such a test, individually caged
hamsters are adapted to a daily schedule of drinking from a single calibrated tube
for one hour at 10 a.m. and again, one hour at 3:30 p.m. with an ad lib food supply.
With rats the morning drinking session is usually only 15 minutes in duration.

On the morning of the conditioning day, the hamster (or rat) is presented with a
solution to be the conditioned stimulus, e.g., sucrose, amonium chloride, acetic
acid, or any of a number of solutions. After five minutes drinking and tasting,
each hamster is given a 30 mg/kg intraperitoneal (IP) injection of apomorphine
hydrochloride. Although the animals do not vomit to this emetic, they show clear
behavioral signs of lassitude and discomfort. They look sick! They recover after
two days and can be tested again with solutions in the morning session. In the
afternoon session, only water is presented to provide adequate hydration for those
animals who show an aversion to drinking the solution presented in the morning
session. A control group receiving water only and no taste also receive IP injec-
tions.

If a given experimental group conditioned to saccharin as the conditioned stimulus
(CS), the degree to which it drinks less of one of the four basic taste solutions,
let us say sucrose, compared to the control group would indicate that the conditioned saccharin aversion has generalized to the sucrose. A percent suppression
score can be derived from the expression:

$$100 \quad x \quad (1 - \text{ml. intake EXP sol.} / \text{ml. intake CON. sol.})$$

If the experimental group drinks an average of 1 cc saccharin and the control group
drinks 4 cc then score is 100 x (1-1/4) = 100 x 3/4 = 75% suppression. Thus a positive score reflects less drinking of the experimental solution. The higher the
score the greater the generalization, and thus the greater the similarity between
conditioned stimulus and the test solution.

The results of such generalization tests of each of the basic taste stimuli against
each other are shown in Figure 9. Each of the basic tastes show significant
suppression scores (i.e. positive generalization to itself). That is, rats poisoned
after sucrose reject sucrose, but not sodium chloride (N), hydrochloric acid (H)
nor quinine hydrochloride (Q). Only HCl gives evidence of cross generalization to
some degree to quinine. These behavioral data may be compared directly to the
electrophysiological response profiles of the chorda tympani and glossopharyngeal
nerves of the rat. Figure 10 shows the means of response frequencies in some
140 units grouped according to their best stimulus categories as sucrose-best,
NaCl-best, HCl-best and quinine-best, as they occur in each of the three gustatory
papillae on the tongue. Thus, although any one taste fiber itself may respond to
more than one basic taste to some degree, when concentrations are used which fall
in the middle of a stimulus effective physiological range, not only is there the
corresponding peak value, but the side responses to other stimuli fall into four
repeating profiles. Units most sensitive to sugar respond second-best to sodium
chloride, third to HCl and very little to quinine. The other units show similar
regularity. The acid-best fibers are more broadly tuned responding second-best
to NaCl, third best to quinine and least best to sucrose. Significant responses
to quinine are seen only in units from the circumvallate and foliate papillae.
Here the order of effectiveness of other stimuli is H > N > S, but the relatively

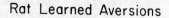

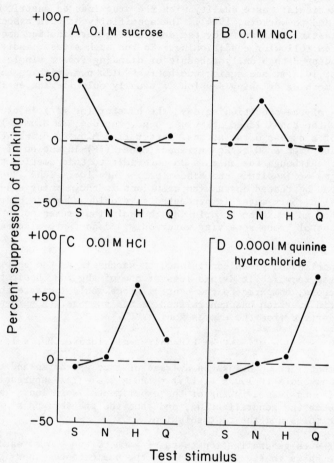

Fig. 9. The mean scores as a measure of generalization in each of four different
         groups of 12 rats. Each group was aversively conditioned to each of the
         stimuli, sucrose, NaCl, HCl, or quinine at the concentrations shown. Each
         group was then tested in sequential daily morning sessions with each of
         the four stimuli including the one with which poisoning had been associated
         The four stimuli have been arbitrarily arranged in order from left to
         right, S, N, H, Q and data points connected with lines merely to facili-
         tate comparisons with the electrophysiological data of Fig. 10 (Nowlis
         and colleagues, 1979).

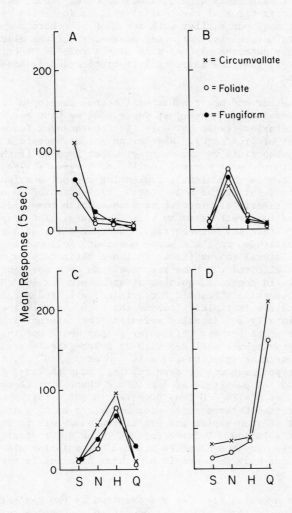

Fig. 10. Composite of rat peripheral taste nerve mean response profiles for 10,11
          and 12 sucrose-best (A), 11, 6 and 25 NaCl-best (B), 12, 13, and 11
          HCl-best (C), and 19, 10, and 0 quinine-best (D) units innervating the
          circumvallate, foliate, and fungiform papillae respectively. Test
          molarities varied from .1 to .5 sucrose (S), .1 to .3 NaCl (N), .001 to
          .01 HCl (H), .001 to .02 Q (from Nowlis and Frank, 1977).

higher frequency of response to H, N and S compared with the other three categories
are 20% or less of the relatively high quinine response. Thus, categorizing affere
sensitivities of taste in terms of the four classic basic tastes seems to be a vali
classification even though our earlier work and that of others seemed to negate thi
relationship. This is not to say that other parameters of the discharge such as
temporal or cross fiber patterns may not also play a role in discrimination of qual
ity. We think, however, that the major quality distinction is based upon these
best stimulus classes.

Let me now turn to another and newer behavioral method developed in our laboratory.
In most of the behavioral data I have just described, we have not looked directly
at taste stimulated behavior itself but only its consequences (except perhaps in
the direct measurement of lick rate). Whether an animal accepts a stimulus or re-
jects it is judged by the fluid volume changes. What happens if the stimulus is
presented directly into the oral cavity? Grill and Norgren (1978a) have developed
the taste reactivity test by chronically implanting two polyethylene tubes, one
into each cheek at the level of the molar teeth. Injection of 50 u liters of
solution produced two clearly distinctive responses which they videotaped to give
a permanent record and to permit frame by frame analysis. The responses proved to
be highly consistent within and between rats. Sucrose elicited a rhythmic sequence
beginning with low amplitude, rhythmic mouth movement, followed by rhythmic tongue
protrusions and then lateral tongue flicks followed ultimately by swallowing. NaCl
and surprisingly, HCl elicited the same response. No body movement accompanied
these oral responses. In contrast, quinine at and above 3×10^{-5} M ($\frac{1}{2}$ log step
above the behavioral aversion threshold for quinine) elicited a response beginning
with gaping. The mandible is rapidly lowered and concomitantly the corners of the
mouth retract posteriorally and dorsally revealing the internal labia. Retraction
of the corners of the mouth forms a triangular shaped mouth opening (see Fig. 11).
Gapes occurred in rhythmic bursts of 2-5 with an inter-gape interval of 85-115
msec. With stronger quinine solutions, 3×10^{-4} M or 3×10^{-3} M, the gapes were
followed by a stereotyped sequence of chin rubbing, head shaking, face washing,
fore limb flailing and paw pushing. At 3×10^{-3} M these body responses occurred
in 80% of the trials, at 3×10^{-4} M they occurred in 30%. Rhythmic bursts of
gaping frequently recurred between successive behaviors of this sequence (See
Fig. 12). The result of such gaping and associated behaviors is to expel and remov
the offending quinine solution. Quinine appears to be a prepotent stimulus for
this aversive response. Why the normally avoided HCl solution did not elicit this
aversive oro-mimetic and body response is something of a puzzle that requires
further study.

The conditioned taste aversion response was examined by the taste reactivity test
(Grill, 1975). A stimulus that normally produces an acceptance response composed
of rhythmical movements of the mandible and tongue and lateral tongue flicks comes
to elicit a replica of the rejection response to quinine after a single pairing
with illness produced by intraperitoneal injection with LiCl. This response includes the full repertoire of gaping, chin rubbing, paw shakes, etc. This altered
taste response persists for approximately one month in the normal rat. This represents one of the purer instances of Pavlovian response substitution as a result
of conditioning.

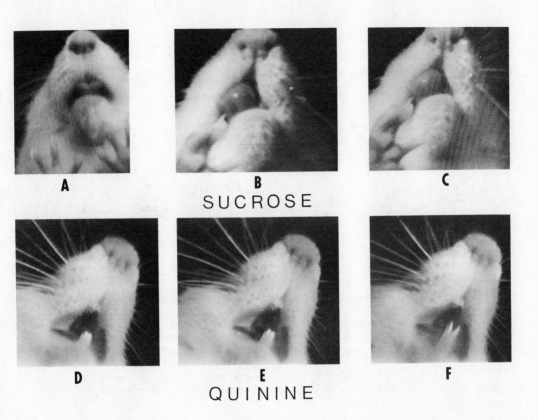

SUCROSE

QUININE

Fig. 11. The stereotyped response sequence by intra-oral
injection of 1.0 M sucrose: (A) tongue protrusions on the midline, (B) and (C) individual lateral
tongue extensions. The response to quinine of
3×10^{-5} M or greater begins with gaping (D-F).
(D), (E), and (F) are sequential frames 16.6 msec.
apart showing opening and retraction of corners of
the mouth and lower lip. (From Pfaffmann, Norgren and Grill, 1977)

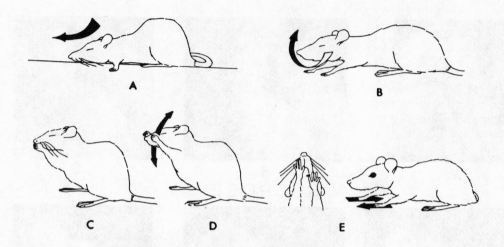

Fig. 12.   The gaping response to strong quinine (3×10^{-4}
and 3×10^{-3} M) is followed by a stereotyped sequence consisting of (A) chin rubbing, (B) head
shaking, (C) face washing, (D) foreleg flailing,
(E) paw pushing, ventral and lateral views. This
sequence appears to facilitate removal of fluid from
the oral cavity and forms part of a total aversive
response pattern. (From Pfaffmann, Norgren and
Grill, 1977).

In the course of re-examining the central nervous system neuroanatomy of taste, Dr.
Norgren and his associates (Norgren and Leonard, 1971; 1973; Norgren, 1974; 1976)
found physiological and anatomical evidence for an important taste relay in the
dorsal pons of the rat. He and I (1975) recorded multiple and single unit electrophysiological activity from this region which indicated that both the anterior and
posterior tongue taste sensitivity were topographically represented. This relay
gives rise not only to a projection to the now well established gustatory thalamus
en route to the cortical projection, it is also the site of a bifurcation in the
ascending pathways from the solitary nucleus of the medulla. The parabrachial nucleus of the pons gives rise to a significant ventral pathway with projections to
the lateral hypothalamus, substantia inominata, central nucleus of the amygdala and
bed nucleus of the striae terminalis as illustrated in Fig. 13. The parabrachial
nucleus is not only a taste relay but receives a significant input from the vagal
afferents which project to the posterior solitary nucleus. It is this whole
system, the special visceral sense of taste plus the general visceral afferents of
the vagus that come together in the parabrachial area to project from there to the
limbic system. It is probably this convergence that mediates some of the special
affinity that taste shows for association with visceral distress. Taste rather than
touch, sight, sound, or other modalities for the rat at least, had a unique pre-
disposition to be associated with visceral distress as the unconditioned stimulus.

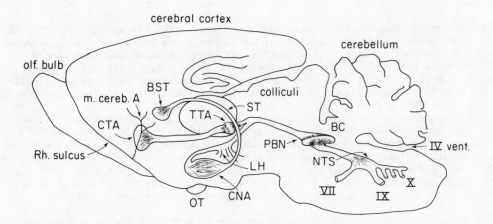

Fig. 13.   Schematic representation of medullary pontine thalamo-cortical and ventral hypothalamic amygdala taste pathways of the rat brain. BC, brachium
conjunctivum; BST, bed nucleus of stria terminalis; CTA, cortical taste
area; CNA, central nucleus of the amygdala; LH, lateral hypothalamus;
m. cereb A, middle cerebral artery; NTS, nucleus tractus solitarius; OT,
optic tract; PBN, parabrachial nucleus; Rh. sulcus, rhinal sulcus; ST,
stria terminalis; TTA, thalamic taste area; IV vent, fourth ventricle;
VII, IX, X, seventh, ninth, and tenth cranial nerve afferents (From
C. Pfaffmann in (Norgren, 1977)).

One question of interest was to know the degree to which this whole ponto-bulbar
system could mediate the oro-mimetic responses just described and indeed, whether
taste avoidance conditioning could be established by it alone. In 1951, Macht had
reported that chronic decerebrate cats could not feed themselves nor select nutri-
ents but that they would swallow when food was placed directly in the mouth. Meat
which was normally swallowed, would, however, be rejected when dipped in bitter
solution, and a sugar impregnated cotton pledged was swallowed. Grill and Norgren
(1978b) were able to prepare chronically decerebrate rats that with careful nursing
were maintained in quite good physical condition with a significant behavioral
repertoire. They maintain a righted posture, groom spontaneously, walk, run, and
jump when stimulated. They do not thermoregulate, and are permanently adipsic and
aphagic. Yet in the taste reactivity test, they executed the same mimetic responses
and very similar motor responses as did normals. Mouth movements, tongue protrusions and lateral tongue movements were elicited by sucrose, HCl and NaCl. The
quinine aversion response like in the normal, was concentration dependent, gapes
appearing only at about the same concentration as in the normal. Chin rubbing,
headshaking and face washing, however, occurred in only 40% of the trials at
3×10^{-3} and were not in evidence at 4×10^{-4} M QHCl. For the normals chin

rubbing was always part of the quinine response at these values. At the lower concentration, head shaking, face washing, forelimb flailing and paw pushing were
regularly seen. In the decerebrates, forelimb flailing and paw pushing did not
occur. Thus, the basic oral aversion response did occur in decerebrates with some
not all, of the more complex body movements. From the point of view of discrimination of good from bad tastes, that capability seems complete within the ponto-bulba
system.

The conclusion that the basic discriminatory and behavioral mechanisms for acceptance or rejection are organized at the relatively low brain stem level in animals
is consistent with Steiner's observations on human neonates who show clearly
differential oral-facial mimetic responses to taste stimuli (1973). Sweet stimuli
caused a marked relaxation of the face and an expression resembling satisfaction
and pleasure, often a slight smile and incipient sucking and licking. Acid solutions caused lip pursing, often accompanied by eye blinking and nose wrinkling.
Quinine caused a typical arched opening of the mouth with upper lips elevated,
jaw depressed and tongue protruded, often followed by spitting or movements
preparatory to vomiting. These gusto-facial responses could be seen even in
anencephalic infants who survive long enough for testing. Post mortems indicated
that they had no functional brain tissue above the ponto-bulbar brain stem. Thus,
the animal and human studies are in agreement in pointing to a complete neural
organization of this relatively "hard wired", most basic response system.

Presumably the more cephalad brain structures including limbic structures play a
role in the motivation and organization of food seeking behavior in accordance
with the dictates of the scale of values of the brain stem (Pfaffmann, Norgren
and Grill, 1977). The more cephalad structures also probably participate in the
modifiability of behavior through learning involved in the conditioned taste
aversion phenonmenon. This, in fact, was the object of a direct test (Grill and
Norgren, 1978c). Five decerebrate and two full surgical control rats were examined
for their capacity to retain the conditioned vaersions acquired before the transection. They were also examined for their capacity to acquire the same association
after transection. The experimental animals were exposed to four daily taste-nausea
pairings before surgery. After the transection there were 8 daily pairings, amounts
of training far in excess of that required in a normal animal. In addition, the
animals were tested under exogenous arousal by tail pinch or amphetamine injection
to insure optimal behavioral arousal. Under none of these conditions did the decerebrate give evidence of retaining a previously learned aversion nor of acquiring
one.

I can best conclude this review which I have organized largely around the work and
interests of our laboratory group, my colleagues Drs. Frank, Nowlis and Norgren and
myself, by paraphrasing the conclusions of our recent review of Taste Neuropsycho-
logy (1979) as to possible future developments and directions of research.

> As has long been known, most single taste afferent units respond
> to a range of chemicals. The "best stimulus" designator or marker
> provides a way of bringing more order into the classification of
> units, especially in relation to certain specificities. For
> example in rats, quinine-best fibers are found almost exclu-
> sively in the glossopharyngeal nerve. The way is now clear
> also to examine the chemical relations among stimulating agents
> within each such class. In the next decade, we can effect final
> resolution of the controversy of a few basic taste receptors
> versus a multireceptored array with the study of a wider range
> beside the classical tastants. More stress will be laid upon
> analytical animal psychophysics on the same species for which
> electrophysiological data exist or can be obtained. The best

stimulus designations of taste units may turn out to be a con-
ceptual as well as a terminological advance.

We can expect many developments on the CNS aspects of taste
as the neural and functional relations of the ventral taste
pathway and its limbic projections are revealed. Increasingly
here, as in the study of peripheral mechanisms, we can expect
exciting new advances that will link gustatory function even
more closely to appetite, food and fluid intake, preferences
and aversions, and hedonic processes. Insights into the control
of behavior by gustatory stimuli may ultimately illuminate the
physiological nature of reinforcement.

ACKNOWLEDGEMENT

Over the years the research of our laboratory has been supported by grants from the
National Science Foundation and the U.S. Public Health Service. This report was
prepared under support of NSF BNS 78-16533, NSF BNS 76-81408 and PHS 10150.

REFERENCES

Andersson, B., S. Landgren, L. Olsson, and Y. Zotterman (1950). Acta Physiol.
   Scand., 21, 105-119.
Brown, J.S. (1955). Pleasure-seeking behavior and the drive-reduction hypothesis.
   Psychol. Rev., 62, 169-179.
Cabanac, M. (1971). Physiological role of pleasure. Science, 173, 1103-1107.
Davis, J.D., and M.W. Levine (1977). A model for the control of ingestion.
   Psychol. Rev.,84, 379-412.
Deerr, N. (1949). The History of Sugar, Vol. 1 and 2. Chapman and Hall, London.
Diamant, H., B. Oakley, L. Strom, C. Wells, and Y. Zotterman (1965). A comparision of
   neural and psychophysical responses to taste stimuli in man. Acta Physiol.
   Scand., 64, 67-74.
Diamant, H., and Y. Zotterman (1959). Has water a specific taste? Nature (Lond.)
   183, 191-192.
Garcia, J., W.G. Hankins, and K.W. Rusiniak (1974). Behavioral regulation of the
   milieu interne in man and rat. Science, 185, 824-831.
Garcia, J., D.J. Kimeldork, and R.A. Koelling (1955). Conditioned aversion to
   saccharin resulting from exposure to gamma radiation. Science, 122, 157-158.
Grill, H.J. (1975). Sucrose as an aversive stimulus. Neuroscience Abst., 1,
   525.
Grill, H.J., and R. Norgren (1978a). The taste reactivity test. I. Mimetic responses to gustatory stimuli in neurologicall normal rats. Brain Res., 143,
   263-279.

Grill, H.J., and R. Norgren (1978b). The taste reactivity test. II. Mimetic re-
   sponses to gustatory stimuli in chronic thalamic and chronic decerebrate rats.
   Brain Res., 143, 281-297.
Grill, H.J. and R. Norgren (1978c). Chronic decerebrate rats demonstrate satiation
   but not baitshyness. Science, 201, 267-269.
Macht, M.B. (1951). Subcortical localization of certain "taste" responses in the
   cat. Fed. Proc., 10, 88.
Norgren, R. (1974). Gustatory afferents to ventral forebrain. Brain Res., 81, 285-
   295.
Norgren, R. (1976). Taste pathways to hypothalamus and amygdala. J. Comp. Neurol.,
   166, 17-30.

Norgren, R., and C. Leonard (1971). Taste pathways in rat brainstem. Science,
    173, 1136-1139.
Norgren, R. and C. Leonard (1973). Ascending central gustatory pathways. J.
    Comp. Neurol., 150, 217-238.
Norgren, R. and C. Pfaffmann (1975). The pontine taste area in the rat. Brain
    Res., 91, 99-117.
Nowlis, G.H., and M. Frank (1977). Qualities in hamster taste: Behavioral and
    neural evidence. In J. LeMagnen and P. MacLeod (Eds.), Olfaction and Taste VI.
    Information Retrieval, London.
Nowlis, G.H., M. Frank, and C. Pfaffmann (1979). Generalization among taste sti-
    muli in rat and hamster (in preparation).
Nowlis, G.H. and W. Kessen (1976). Human newborns differentiate differing con-
    centrations of sucrose and glucose. Science, 191, 865-655.
Papousek, H. (1967). Experimental studies of appetional behavior in human new-
    borns and infants. In H.W. Stevenson, E.H. Hess and H.L. Rheingold (Eds.),
    Early Behavior :   Comparative and Developmental Approaches. Wiley, New York,
    pp. 249-277.
Pfaffmann, C. (1939). Specific gustatory impulses. J. Physiol. 96, 41-42.
Pfaffmann, C. (1960). The pleasures of sensation. Psychol. Rev.. 67, 253-268.
Pfaffmann, C. (1969). Taste preference and reinforcement. In J. Tapp (Ed.),
    Reinforcement and Behavior. Academic Press, New York. pp. 215-240.
Pfaffmann, C. (1977). Biological and behavioral substrates of the sweet tooth.
    In J.M. Weiffenbách (Ed.), Taste and Development: The Genesis of Sweet
    Preference. DHEW publ. no. (NIH) 77-1068, Bethesda. pp. 3-24.
Pfaffmann, C. The vertebrate phylogeny of taste, neural code and integrative pro-
    cesses. In E.C. Carterette, and M. Friedman (Eds.), Handbook of Perception,
    Vol. 6. Academic Press, New York. pp. 51-123.
Pfaffmann, C., M. Frank, and R. Norgren (1979). Neural mechanisms and behavioral
    aspects of taste. Ann. Rev. Psychol.,30, 283-325.
Pfaffmann, C. and E.C. Hagstrom (1955). Factors influencing taste sensitivity to
    sugar. Amer. J. Physiol., 183, 651.
Pfaffmann, C., R. Norgren, and H.J. Grill (1977). Sensory affect and motivation.
    Ann. NY Acad. Sci.,290, 18-34.
Richter, C.P. (1942). Self-regulatory functions. Harvey Lectures, 38, 63-103.
Rozin, P. (1978). The use of characteristic flavorings in human culinary prac-
    tice. In C.M. Apt (Ed.), Flavor: Its Chemical, Behavioral and Commercial
    Aspects. Westview Press, Boulder, Colorado. pp. 101-127.
Steiner, J.E. (1973). The gusto-facial response: observation on normal and
    anencephalic newborn infants. In J.F. Bosmi (Ed.), Fourth Symposium on Oral
    Sensation and Perception. Washington Superintendent of Documents, U.S. Govern-
    ment Printing Office, Washington. pp. 254-278.

STRUCTURAL CHANGES IN THE EXCITABLE MEMBRANE DURING EXCITATION

A. von Muralt

Theodor Kochor Institute, University of Bern, Switzerland

ABSTRACT

The excitable membrane in nerves permits the registration of three separate signals during excitation: the electrical spike, the optical spike and the thermal spike. All three are the signs of a high sensitivity of the membrane towards depolarisation and repolarisation, caused by structural changes within the membrane.

KEYWORDS

Reversible changes of birefringence and heat production and absorption in the excitable membrane; optical spike; thermal spike; structural changes in the excitable membrane.

INTRODUCTION

The great teacher of physics in my student days at the university of Zürich was Erwin Schrödinger. Seventeen years later he gave a remarkable series of lectures at Trinity College in Dublin, with the title "What is Life?" (Schrödinger, 1944). In these lectures he showed that all living beings maintain Life on the basis of a high internal "orderliness" and this in an environment where every transformation of energy leads invariably to a partial "randomization" (Second law of thermodynamics). Heat, which is molecular disorder D is the endpoint of this universal tendency and its reciprocal $1/D$ can be used as a measure for molecular order. Since the logarithm of $1/D$ has a negative sign, one can write the Boltzmann equation in the following way:

$$-(\text{Entropy}) = k \cdot \log 1/D \quad (k = \text{Boltzmann constant})$$

with other words: Entropy with a negative sign is in itself a measure of orderliness! All polysaccharides, proteins and lipids are molecules of a high degree of internal orderliness and their free energy is available for the maintenance of the high degree of organisation in all living beings against the dominating tendency towards "randomization". Schrödinger expressed this unique faculty of living beings by saying: "They feed upon negative entropy."

Is it possible to register rapid and reversible changes of "orderliness" in such living tissues as nerves or muscles? The answer is yes - but only under certain conditions. Two, seemingly quite different physical approaches can give definite informations: the registration of thermal changes during activity and the registration of variations of birefringence! This sounds strange, so a simple model, taken from physical chemistry may serve as an explanation for this somewhat surprising interrelation of two entirely different approaches. If water freezes into ice it changes from "randomization" into crystalline order, the entropy decreases and heat is produced. The ice crystals show birefringence as a sign of their internal structure. When the ice melts the birefringence disappears and heat is reabsorbed by the system. Heat <u>production</u> accompanies the transition into a higher degree of "orderliness" and heat <u>absorption</u> comes with the disappearance of this state into "randomization". Birefringence is the optical sign of "orderliness" and disappears when the ice is melting. This simple model may help the reader to understand the relation between heat and birefringence.

Let us now switch over to the basic phenomena, which happen in the excitable membranes of nerves during excitation. 25 years ago Alan Hodgkin and Andrew Huxley made a very decisive step, deepening our knowledge by a mathematical treatment of the ionic events, which are the basis of the action potential in nerves. The excitable membrane has a unique property, it is highly sensitive to potential changes and responds to small depolarisations in such a way, that the energy stored in the unequal distribution of ions - sodium high outside and potassium high inside - is used for the creation of the action potential, which is the unique "signal" in all nerves. A small depolarisation of the resting potential in the excitable membrane opens the sodium channels in this membrane and sodium ions flow inward, thus increasing the depolarisation even to an "overshoot" to the positive side. With a delay the potassium ions begin to flow outward, about 10 times slower than the sodium ions, thus repolarising the membrane potential to its resting value. The discovery of chemical agents, which are able to "plug" the ionic channels led to two important discoveries: 1. There are two kinds of ionic channels, sodium- and potassium-channels. 2. A depolarisation of the excitable membrane leads primarily to a "gating current" as a result of the movement of charged subunits within the sodium channel.

The chemicals which "plug" selectively the channels are Tetrodotoxin (TTX) or Saxitoxin (STX) for the sodium channels, "plugging" them from the outside and Tetraethylammonium (TEA) which acts only from the inside, "plugging" the potassium channels. With the aid of TTX and STX the number of channels per square µ in a nerve can be determined and Table 1 shows in a very clear way how very small the area of the sum of sodium channels is, compared to the surface of the excitable membrane!

TABLE 1   Surface Relations of Na-Channels and Membrane

Species	Squid giant axon	Rabbit vagus n.	Pike olfactory n.
Nr. of Na-channels	$500/\mu^2$	$25/\mu^2$	$3/\mu^2$
Surface $\overset{o}{A}{}^2$	500 x 19.63	25 x 19.63	3 x 19.63
Channel Surface / Membrane Surface	$9815 : 10^8$	$491 : 10^8$	$59 : 10^8$

This leads us to the very important question: is the excitable membrane, with its high potential-sensitivity only a carrier of sodium and potassium channels and a passive transmitter of potential changes or is the excitable membrane also active as a partner in the creation of the "signal", the action potential?

In recent years new researches have shown that remarkable and rapidly reversible structural changes occur in the excitable membrane, synchronous with the action potential.

Birefringence in Nerves

In physics teaching nobody cares much about the birefringence in biological objects and in biology teaching there are only few teachers who know enough about the basic laws of birefringence. A short survey may therefore be helpful.

Birefringence in biological objects is always a sign of organized macromolecular structures. The optical axis, which is the symmetry axis of the refractive power of an object, coincides in muscles with the fibre axis, the refractive index, measured parallel to the fibre axis, $n_\parallel$, is larger than the refractive index $n_\perp$, perpendicular to it, so the plotted spatial distribution of refractive indices is not a sphere, but an ellipsoid. The difference $n_\parallel - n_\perp$ is a measure of the size of birefringence in a given fibre. In nerves, such as the giant axon of the squid there is a large axon-cylinder, surrounded by a relatively thin axon-membrane, containing the excitable membrane. In the axon the optical axis coincides with the fibre axis, but in the axon-membrane, where the interesting optical changes occur, the optical axis is a radial-corona enclosing the axon over 360^0. Viewed under a microscope the optical plane, focussed on the center of the fibre, cuts out a horizontal portion of the axon and an equal portion of the limiting "corona" on both sides. In the "corona" $n_\parallel - n_\perp$, with respect to the fibre axis is negative!

The classical method of measuring birefringence makes use of a calibrated compensator, which is inserted into the light path with its optical axis at right angles to the optical axis of the specimen and shifted in such a way until the retardation R_C of the compensator is equal and opposite in sign to the retardation R_O of the object. At the point of maximal darkness $R_C = R_O$.

For the study of very rapid changes in the biological object this method cannot be used, but the change of birefringence is always accompanied by a change in brightness ΔI of the outcoming elliptically polarised light - and this change can be registered by photocells, coupled to an amplifier and an oscillograph. The retardation R between the two components of the elliptically polarised outcoming light is related to the differences of the refractive indices by

$$n_\parallel - n_\perp = \frac{R}{d}$$

where R is the retardation and d the length of the light path in the birefringent object. If there is a rapid and reversible change in the retardation ΔR the light intensity changes by an amount of ΔI and the calculation shows for very small intensity changes ΔI that the change in retardation ΔR is

$$\Delta R = \frac{1}{2}(\frac{\Delta I}{I}) \cdot R$$

(For the full development of these formulas see von Muralt, Weibel & Howarth, 1976)

BIREFRINGENCE CHANGE DURING EXCITATION

Transient, extremely small changes of birefringence during the passage of an action potential in nerves were discovered in 1968 and 1969. The pioneers in these studies were L.B. Cohen and R. Keynes (1968), Tasaki, Watanabe, Sandlin and Carnay (1968), Entine (1969) and Berestovsky, Lunevsky, Razhin and Musienko (1969) using crab nerve, lobster nerve, crayfish nerve, rabbit vagal nerve and mainly squid giant axons. Cohen, Hille and Keynes (1970) and Cohen, Hille, Keynes, Landowne and Rojas (1971) found that the transient optical change closely follows the time course of the intracellularly recorded action potential in the squid giant axon. I had the rare pleasure to participate during two "squid seasons" in the remarkable experiments of Cohen, Keynes and Rojas in Plymouth and I remember vividly how we sat in a vibration-proof cellar, watching during the hours after midnight the slowly growing birefringence signal, rising with each excitation wave out of the "noise", fascinated to see this optical "witness" of transient structural changes in the excitable membrane. Thousands of impulses had to be stored in an averaging computer in order to get a reliable document of this "optical spike".

On my return to Bern I decided to try another nerve, the olfactory nerve of the pike, Esox lucius, with which I was familiar from some previous work. This nerve is remarkable in many ways: it is built up by 4 million extremely small nerve fibres with a modal diameter of 0.2 µ! They can only be seen with the aid of the electron-microscope and each axon is bounded by an axon-membrane (excitable membrane) forming a bilayer about 8 nm thick, surrounding the axon with a diameter of about 180 nm. The axon contains 4-20 microfilaments and 2-5 microtubules of about 25 nm diameter. These microtubules and microfilaments assume a straight course along the nerve fibre, which has a length of 50-70 mm, depending on the size of the pike. 90 % of all fibres range in diameter from 0.1 - 0.5 µm, with a high peak of the distribution curve at 0.19 µm and the number of fibres in small olfactory nerves is $4 \cdot 10^6$ going up to $20 \cdot 10^6$ in large nerves. 1 mm^3 of olfactory nerve contains 8400 mm^2 of axonal, excitable membrane! So, in working with this nerve there is no need for an averaging computer - the computer is "built" into these millions of nerve fibres! I obtained a first transient, rapid and reversible change of birefringence from one nervous impulse in September 1970 - and my joy was great! There was the real optical spike! Figure 1 shows a registration of an optical spike (change of birefringence) OS during the passage of one excitation wave, registered by its action potential AP.

Fig. 1

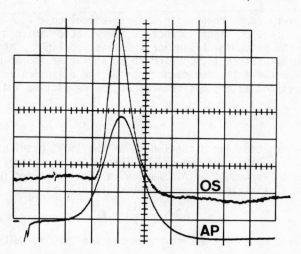

The optical spike OS was registered with an oscillograph during the passage of one nervous impulse, together with the electrical change, the action potential AP. The temperature was 0° C and the maximal change of retardation in the optical spike is $\Delta R = 0.05$ nm, the peak of the action potential AP = 3.8 mV and the time scale 100 msec/division.

This increase in birefringence in the excitable membranes of the 4 million nerve fibres must be interpreted as a macro-molecular "tightening" of structure around the axons, or with other words an increase of "orderliness" within the structure of the excitable membrane! This rapid structural change is fully reversible and its release into the resting state follows exactly the time course of the action potential, even in the transient hyperpolarized phase of the action potential. In which way these structural changes act on the opening and closing of the sodium- and potassium-channels remains a problem open for further studies.

THE THERMAL SPIKE

In all nerves each transmission of signals is followed by an increase of their oxygen consumption, their glycogenolysis and of their thermal output (recovery heat) up and above the resting heat production. The primary source of energy for these recovery processes is the hydrolysis of ATP (sometimes also called the Angel of Terrestrial Planning). The free energy of ATP drives the ion-pump in the excitable membrane, which transports sodium ions "uphill" against the concentration gradient back to the outside of the membrane and potassium ions also "uphill" back into the inside - something like a revolving door in a hotel-entrance. This metabolic activity creates heat, the so-called "recovery heat". Recent careful measurements of Howarth, Ritchie and Stagg (1979) have shown that this heat is 381 ± 26 µcal/g · impulse at 20° C in the olfactory nerve of the garfish. The recovery heat corresponds to the increased oxygen uptake of the nerve after activity.

During activity, however, there is also an outburst of heat, the "initial heat" of A.V. Hill, which occurs during the rising phase of the action potential. The accurate analysis with a highly sensitive thermopile assembly, constructed by the master in this field, Victor Howarth, revealed that the passage of an action potential releases a small quantity of heat, simultaneously with the rising phase of the action potential; this heat is <u>completely</u> reabsorbed during the falling phase of the action potential. So there is also a "thermal spike". Howarth, Keynes and Ritchie (1968) had found in rabbit non-myelinated fibres that the positive heat from the rising phase of the action potential was three times greater than the most generous estimate based on a free energy exchange from metabolism. So they suggested that the excitable membrane becomes more ordered thermodynamically, with other words there is a decrease of entropy, a decrease which is characteristic for water on freezing into ice. The olfactory nerve of the pike has the great advantage that its 4 million fibres conduct the excitation wave at the same speed (practically) and with a new very small and highly sensitive thermopile Victor Howarth, Richard Keynes and Murdoch Ritchie came to my laboratory in Bern, where we were lucky enough to register a "thermal spike" during the passage of the action potential in the olfactory nerve of the pike. Among the younger membres of the staff we were called "The four old foxes".

Fig. 2

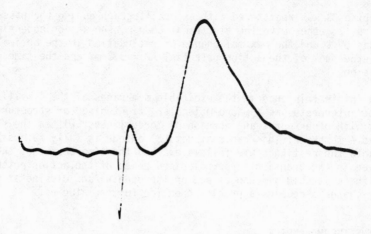

.Figure 2 shows this "thermal spike", originating during the passage of the nervous impulse in the olfactory nerve of the pike (Howarth, Keynes, Ritchie and von Muralt, 1975). The record of Fig. 2 was obtained by averaging 20 impulses. The duration and the rise and fall of the temperature curve are synchronous with the action potential, but the real phases of heat production and heat reabsorption, which is a little greater than the production, has to be determined by a heat block analysis. This analysis shows that the corrected heat spike follows exactly the time course of the action potential.

What is the explanation of these rapid changes from heat production to heat re-absorption during one nervous impulse? There are several theoretical approaches: 1. One can consider the excitable membrane as a charged condenser which discharges through the sodium inflow and recharges with the potassium outflow. The calcula-tion shows that such a condenser-discharge and recharge could account at the most for 1/5 of the heat production and reabsorption. The optical spike, with its in-crease and decrease of birefringence shows beyond doubt that there is an increase of "orderliness" in the excitable membrane - thus a decrease of entropy - in the rising phase and a reversal in the falling phase of the action potential. We wrote in 1975: "On general grounds, changes in entropy of the phospholipid nerve mem-brane would be expected to accompany the large changes in voltage gradient, as-sociated with the action potential." Since then new studies on the thermal spike in garfish olfactory nerve fibres by Howarth, Ritchie and Stagg (1979) have shown that the thermal spike is larger in garfish than in the pike and permits a much more accurate analysis of the time relations between the rising and the falling phase of the thermal spike. They came to the conclusion that "entropy changes make a significant contribution to the observed heat change", and they stated that "free energy changes and entropy changes together can account for all observed heat, assuming credible values for the membrane potential and action potential".

The structural changes, lowering and rising the entropy of the excitable membrane must give also other measurable signs. There are the very interesting studies of rapid changes of fluorescence during excitation ("fluorescence spikes") if the membrane has been stained with various special dyes. Larry Cohen and Tasaki are the leaders in this field, of which I gave a summary at a symposium of the Royal Society (von Muralt, 1975). And then there seem to be even extremely small, reversible changes of the thickness of the excited axon, of the order of 18 Å during the passage of the impulse (Hill, Michelson, Nokes and Schubert, 1977).

Our great friend Yngve Zotterman has been a pioneer in neurophysiology by discovering, more than 50 years ago, together with Lord Adrian that the spikes of the action potential carry in their relative spacing the message of the intensity of the stimulus. At that time the techniques of research in neurophysiology were very primitive and it took a great deal of manual skill and intelligent thought, if one wanted to get straight-forward results. Today there are technical possibilities available of which we did not dare to dream 50 years ago, but the laboratory work is still going on with the same basic spirit, which has animated Yngve's life-work during half a century: Veritas!

REFERENCES

Berestovsky, G. N., V. Z. Lunevsky, V. D. Razhin and V. S. Musienko (1969). Rapid changes in birefringence of the nerve fibre membrane during excitation (in Russian). Dokl. Akad. Nauk. USSR, 189, 203-206.

Cohen, L. B. and R. D. Keynes (1968). Evidence for structural changes during the action potential in nerves from the walking leg of Maia squinado. J. Physiol. Lond., 194, 85-86.

Cohen, L. B., B. Hille and R. D. Keynes (1970). Changes in axon birefringence during the action potential. J. Physiol. Lond., 211, 495-515.

Cohen, L. B., B. Hille, R.D. Keynes, D. Landowne and E. Rojas (1971). Analysis of the potential-dependent changes in optical retardation in the squid giant axon. J. Physiol. Lond., 218, 205-237.

Entine, G. (1969). Physical probes of nerve membrane structure. Thesis, Berkeley, Univ. of California.

Hill, B. C., R. P. Michelson, M. A. Nokes and E. D. Schubert (1977). Laser interferometer measurement of changes in crayfish axon diameter concurrent with action potential. Science, 196, 426-428.

Howarth, J. V., R. D. Keynes and J. M. Ritchie (1968). The origin of the initial heat associated with a single impulse in mammalian mon-myelinated nerve fibres. J. Physiol. Lond., 194, 745-793.

Howarth, J. V., R. D. Keynes, J. M. Ritchie and A. von Muralt (1975). The heat production associated with the passage of a single impulse in pike olfactory nerve fibres. J. Physiol. Lond., 249, 349-368.

Howarth, J. V., J. M. Ritchie and D. Stagg (1979). The initial heat production in garfish olfactory nerve fibres. Accepted for print by Proc. R. Soc. Lond. B.

von Muralt, A. (1975). The optical spike. Phil. Trans. R. Soc. Lond. B, 270, 411-423.

von Muralt, A., E. Weibel and J. V. Howarth (1976). The optical spike. Structure of the olfactory nerve of pike and rapid birefringence changes during excitation. Pflügers Arch., 367, 67-76.

Schrödinger, E. (1944). What is Life? Cambridge University Press.

Tasaki, T., A. Watanabe, R. Sandlin and L. Carnay (1968). Changes in fluorescence, turbidity and birefringence associated with nerve excitation. Proc. natn. Acad. Sci. USA, 61, 883-888.

DISCUSSION

Editorial Note

After the formal presentations by Yngve Zotterman's friends and collaborators, Richard Adrian chaired the Discussion on 'Future Trends', which began with some challenging remarks by Yngve Zotterman. These were taken up and developed by the speakers. There is always an editorial problem with Discussions at symposia. In this instance, a tape recording was made of all the contributions and from this a transcript was prepared. I felt that were this to appear verbatim in print, every contributor ought to have been given the opportunity to edit his comments and in my limited experience this takes a long time and generally results in watering down of exciting and outrageous statements to a tasteless end-product garnished with phrases such as 'I don't think so', 'generally speaking' or 'you may be right'. So Yngve and I decided that I would write a summary of the Discussion, using the taped transcript as the text, apologising in advance for any misinterpretations which may appear. Yngve's introductory remarks are set out verbatim.

Declan Anderson

Introduction

In this Symposium held in my honour and attended by many of my old friends and collaborators, we have learned about the most recent advances in many branches of physiology - about C-fibres, touch and pain, the functions of peripheral and central sensory neurones and transduction in the receptors, etc., the mandibular reflexes and finally Alex von Muralt's exciting report about structural changes during excitation, and we saw how the relatively new field of Oral Physiology has advanced. Since I had the privilege of opening the first international symposium in Oral Physiology in 1971 in the Wenner Gren Centre in Stockholm, this field of research has developed very impressively as you know. An International Commission of Oral Physiology has been established under the auspices I.U.P.S. and a number of creative meetings have been held, some of the most successful here in Bristol under the excellent and devoted leadership of our friend Declan Anderson, who is the secretary of the Commission of Oral Physiology and is Professor here in Oral Biology.

"How come" as my American grandson says - how come that I, originally a neurophysiologist, became interested in Oral Physiology? That is very simple; I found that besides its important motor function in speech, etc., the tongue fulfils so many fascinating sensory functions which go far back in the phylogenetic history of life. These functions are developed very early - lip activity, etc., in all animals. Immediately after birth the infant searches his mother's breast and expresses his delight or refuses artificial feeding which does not satisfy his taste. The lips, the tongue, the gums, the throat are so richly supplied with sensory mechanisms providing protection against non-acceptable or damaging agents while other sensory mechanisms aid the choice of foodstuffs. They also provide the first intimate contact between individuals. It is not my intention to summarize the beautiful contributions made during the meeting. Instead, I would challenge you to what the Germans call "eine Aufforderung zum Tanz", which I make by presenting to my colleagues a problem which has occupied my thoughts ever since Lloyd Beidler came to Stockholm in 1962 to the first Symposium on Olfaction and Taste. He had discovered that the taste buds and the gustatory cells have a very short lifespan, being steadily renewed by ordinary epithelial cells growing in and undergoing a metamorphosis into specific cells with the microvilli etc., and displaying, as we believe, their specific response pattern. We know now that the taste buds disappear when the nerves containing gustatory fibres such as the chorda tympani are severed and we know that the taste buds reappear when the taste fibres which have cell bodies in the geniculate ganglion grow out to the periphery again. We know that the outgrowing of taste fibres can only produce taste buds in regions where taste buds are normally to be found.

I have suggested for at least 12 years to young people in the field that they cross over gustatory fibres to the lips in the monkey where there aren't taste buds naturally. I hope that somebody will do this. I think that there are people who have implanted gustatory neurones in the eye where they can grow, but they have not seen any taste buds. Others maintain that if you give some hormones you can develop taste buds in regions where they do not naturally occur by growing gustatory fibres into these regions. I would like to see that properly controlled and I have had long discussions with Paul Weiss about this

when he was in Stockholm in 1962 and subsequently. He was absolutely sure that
the epithelial cells at the periphery which develop into taste buds are those with
the genetic code which responds to the specific gustatory fibres and I am very
much in favour of that idea but it could be proven if one did the experiment on
the monkey's lips, to which I referred already.

It looks to me as if there are specific agents in the peripheral
neurones which produces structural changes in the peripheral cells. Let me
suppose it is some kind of a peptide. That peptide, I imagine, can be
transported to the nerve terminals by axonal transport where it can exert its
influence on the cells which have the genetic code necessary to enable them to
respond by undergoing the metamorphosis into taste buds. Is it a peptide? I
believe that this is a phenomenon not limited simply to taste buds. I think
that exactly the same thing occurs with sensory neurones in the skin, even in
such structures as the Pacinian corpuscles, Ruffinis, etc. In the genetic code
there must be something which will transform suitable cells in the periphery into
receptor organs like the taste buds. Has it been shown that Pacinian corpuscles
disappear when the nerve supply is sectioned and reappears when the tissue is
reinnervated?

I must tell you that we have made a very preliminary study to look into
this question and I asked Tomas Hokfelt who is an expert on immuno-histochemistry
and can study 2 peptides, one of them Substance P, discovered in the thirties by
my friend Ulf von Euler and which accumulates, as Hokfelt and others have shown,
in the smallest dorsal root neurones. Many believe that this is a transmitter.
I would rather believe it is a moderator to do with specificity. So I have asked
Hokfelt to see what peptide he could find with radio immuno-assay in the
geniculate ganglion. It would be nice if the gustatory cells contained some
specific peptides. To solve these problems we have to rely upon team work
between neurobiologists, sensory physiologists and biochemists.

Just recently, Hokfelt and his co-workers at Karolinska Institutet have
found Substance P in the peripheral endings of gustatory fibres in the
circumvallate papillae of the cat's tongue. They and others have previously
found Substance P in the smallest sensory neurones in the posterior root ganglion
and in the dental pulp - i.e. most likely in specific nociceptive neurones.
Whatever may be the function of Substance P as a transmitter or a moderator, it
cannot be responsible for inducing a specific influence on the peripheral
receptor, its structure and sensory function, unless we consider the possibility
that it may appear with a different structure, e.g. that the sequence of the
eleven amino-acids may change or that their number may vary. In such a case the
variability would be great enough to procure the necessary number of specific
agents for all sensory kinds of specific receptors or sites in the tongue as well
as in the skin. But of course it is perhaps equally likely that the specificity
is given by other substances or peptides which are transported down in the sensory
fibres to the periphery and there induce the specific response pattern of the
receptors.

                                                    Yngve Zotterman

Summary

The first comment in response to Yngve Zotterman's challenge came from Lloyd Beidler who pointed out that no one had yet found specific peptides in taste cells, although they had been found in the olfactory nerve by means of immuno-assay methods in the nerve during regeneration. He reminded us of work he had done a long time ago and had presented at the Munich I.U.P.S. Congress. Tritiated leucine was put into the petrosal ganglion and the taste buds were examined later using autoradiography. Between 12 and 24 hours after injection the taste buds were densely labelled, but one could not be certain whether the label was associated with the nerve fibres or taste cells. It almost looked as though it stayed in the nerve but it would take a great deal of effort to show that it goes into the taste cells and even if radioactivity were present it might be in amino-acids produced by breakdown of the labelled protein and transferred from the nerve terminal to the taste cell and resynthesised.

He then went on to describe experiments performed by one of his students on transport in the very long olfactory nerve of the garfish (Gross, 1973; Gross and Beidler, 1975). Tritiated leucine was applied to the olfactory capsule, and Lloyd showed results which demonstrated that the rate of transport is obviously highly dependent on temperature. Extrapolation of the function to $37^{\circ}C$ reveals a rate of about 400 mm/day which is very similar to that recorded in mammals (Ochs, 1972). Another of his students (Raderman-Little, 1979) had studied turnover of catfish tastebuds as a function of temperature. Lloyd suggested that experiments of this type could be done in the chorda tympani by examining axoplasmic transport between the geniculate ganglion and the taste buds. If this problem could be solved it would provide very important information in biology as a whole.

Yngve then asked Ainsley Iggo to tell us what happens in the skin and Ainsley discussed his preparation using the cat Merkel cell which has the advantage that it can be seen from the surface. If the nerve is cut, the peripheral part degenerates, the Merkel discs disappear and the Merkel cell itself rapidly becomes disorganised and it too disappears after a while. He indicated that there is conflict in the literature as to whether this happens in all species or not, but if the nerve is crushed instead of being cut and is allowed to grow back, the growing tip has an unspecialised sensitivity while the axon is growing back. It does not regain the typical response characteristics until the Merkel cell complex is reformed. He added that there is no information available concerning the appearance of the cells during the intermediate stage. However, he was sure that the special properties of the afferent unit are recovered when the Merkel cell and the nerve ending have got together again, but not until then.

Richard Adian wondered if the Merkel cell was the original one that recovered after reinnervation, or was some other cell transformed. Ainsley replied that this was a question which he was trying to answer using the technique Lloyd Beidler had referred to, but his experiments were as yet in a very early stage.

Ove Franzén offered some comments on sutured nerves. In human patients he had carried out psychophysical studies after the nerve had been allowed some months or longer for regeneration. He said that when vibration is applied to skin, the sensitivity maximum is about 250 Hz, and in these patients he had found that the threshold functions looked normal. The threshold functions for the Pacinian corpuscles parallel the psychophysical functions and he concluded that the Pacinian corpuscle had regenerated and restored its response characteristics.

Andrew Quilliam made a number of comments about Pacinian corpuscle reinnervation. After crushing the splanchnic nerve of the cat, the nerve fibres supplying the mesenteric Pacinian corpuscles degenerate (Quilliam, 1966) but no great functional deficit in the animal's behaviour can be detected by casual observation. However, the capsule of such a corpuscle, its outer core lamellae and the paired stacks of inner core hemi-lamellae remain intact for a considerable period, neither degenerating nor being involved in cellular proliferation (see Figs. 1 & 2).

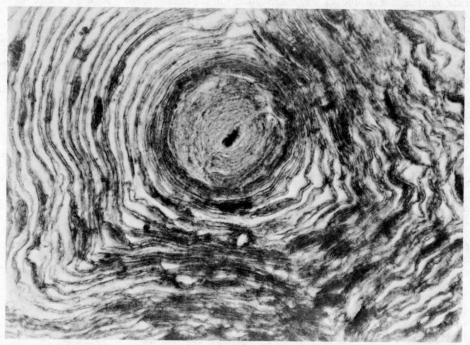

Fig. 1.   Transverse section through a normal Pacinian corpuscle from the
           mesentery of the cat (Silver stain. x 1500).
Although there has been some fixative-induced distortion of the circumferential lamellae of the outer core, the two bilaterally symmetrical stacks of hemi lamellae that constitute the inner core are particularly well depicted. Located axially is the nerve terminal which is somewhat oval in cross section and which is somewhat granular in appearance due to the presence, peripherally, of many mitochondria. Pease and Quilliam (1957).

The numerical stability of the cells involved tends to suggest that conventional Schwann cells are not involved in the construction of most - if not all - of the corpuscle, since elsewhere in the peripheral nervous system, Schwann cells always proliferate when the nerve fibre with which they are associated circumferentially, degenerates.

In the great majority of normal Pacinian corpuscles, there is but a single unmyelinated nerve terminal within the inner core (Quilliam, 1975). Fine pericapsular nerve fibres akin to the "apparatus of Timofeew" can be identified sometimes, but their function remains obscure. Evidently the appearance of a second nerve fibre within the inner core of the feline mesenteric Pacinian corpuscles (as suggested by Santini, 1969) is either a rarity or due to an artefact. On the other hand, tortuosity and/or duplication of the nerve

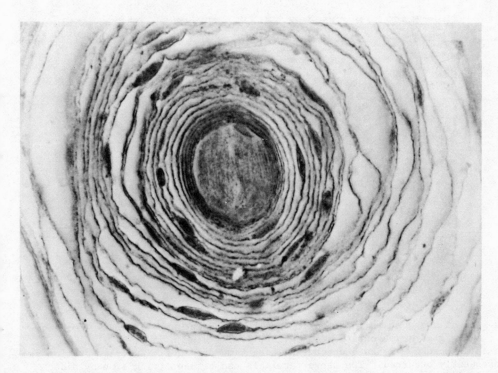

Fig. 2.   Transverse section through a Pacinian corpuscle from the mesentery of a
           cat in which the splanchnic nerve has been crushed. (Silver stain. x
           1500).
The concentric lamellae of the outer core and the two bilaterally symmetrical
stacks of hemi lamellae are clearly depicted.   The empty radial clefts separat-
ing the two stacks of hemi lamellae and a small amount of cellular debris from
the recently degenerated nerve terminal, axially, can be distinguished. No
marked cellular proliferation or degeneration of the concentric lamellae or of
the hemi-lamellae can be detected.

Fig. 3.   A somewhat oblique longitudinal section through a Pacinian corpuscle
            from the mesentery of a cat. (Silver stain. x 750).
The splanchnic nerve has been crushed some time previously and regeneration has
subsequently occurred. Two nerve terminals can be seen lying side by side
axially in the centre of the inner core. Several other unusual features of these
nerve terminals have been noted at other levels in the corpuscle.

terminal can be the occasional outcome of reinnervation of a Pacinian corpuscle subsequent to its previous denervation for any reason (see Fig. 3).

Manfred Zimmermann suggested that so far, the discussion had pointed at a broad heading for a possibly very productive line of research which he thought should be Developmental Neurobiology.He posed various questions in this field. For example, how do sensory nerve cells find their way to the right epidermal cells in the periphery? He felt that this would be a question for research in development in vivo, in post-natal and pre-natal life. Another question was, what are the conditions in which wrong connections are established after nerve lesions? Clinical experiments showed clearly that many wrong connections are possible between nerve cells or between nerve cells and peripheral receptor cells like Merkel cells, and the conditions ought to be investigated. Some fibres which had been connected to a Merkel cell might, for example, after experimental nerve lesions, connect to a Meissner or Pacinian corpuscle. He emphasized the importance of trying to understand the basic mechanisms of cell recognition. Another question which he posed was, how can wrong connections be adjusted in the CNS? This question has been investigated so far mainly in the motor system when nerves to muscles have been exchanged by surgical means.

He did not believe that these questions had been systematically investigated and he thought that the answers would teach us a great deal both of basic scientific importance and also of practical importance such as in repair. It is necessary to find the right preparations for this type of research and in his view the work should be done not only on regenerating nerves after experimental nerve lesions, but should also extend to early life, even embryonic life, and also in tissue culture of the nervous system. He emphasized that this is a line of research which is just coming up at the moment, even using human nerve cells, and he hoped that correlated research by biochemists, anatomists, et al, as Yngve Zotterman had suggested in his opening remarks, would help towards answering not only the questions he had put forward but many others in the future.

Ove Franzén then returned to the Pacinian system, referring to experiments he had performed on patients with unilateral parietal lobe lesions. He found that when tactile stimuli are applied to the skin the threshold on the contralateral side is raised to 10-40 times the normal, and recalled data on the ability of human subjects to match high frequency vibration of up to 384 Hz applied to the finger tips. He had predicted from his results that there is no frequency-following by cortical units above 100 Hz and therefore there can be no representation in single cortical neurones for frequencies higher than 100 Hz. So his conclusion was that this kind of analysis is carried out by subcortical mechanisms. He had predicted that patients with parietal lobe lesions would have unaffected vibratory thresholds and this was indeed found to be so. So with one kind of mechanical stimulation the threshold is tremendously increased but for another kind of stimulation there is almost no effect at all. One patient with a lesion of the left cortex was asked to make estimates of the intensity of tactile pulses within a certain range, applied contralaterally to the lesion. The threshold was tremendously increased.

In the light of his findings, Ove Franzén suggested that conscious experience could be located in other levels than the cortex.

George Gordon referred to the pathways which the Pacinian corpuscles project into and suggested that the system Ove was talking about is one which is clearly the Pacinian system as judged from the tuning curves he had shown. From animal evidence it appears that the dorsal column is the favoured if not the exclusive path that these fibres take and certainly the 2nd order cells respond as if they followed the Pacinian corpuscle frequencies. He mentioned clinical evidence on the sense of vibration, but because the frequency of stimulation is often not specified it is not possible to tell if you are dealing with Pacinian corpuscle or with Meissner corpuscles which are held to be responsible for the fluttering sensation below 100 Hz.

As a clinical neurologist, Peter Nathan then entered the discussion saying that he now believed he had tested vibration for most of his life wrongly and had only realised this in the last few years. His technique was the normal method of a clinician, which he now felt was wrong because when a vibrating fork was placed on a limb, it was in fact testing afferents from bone, joints and deep tissues. From a physiological point of view he thought that this was useless, and he felt that, though rather late in life, he had to start picking up skin and applying vibration to it, or better still to apply the vibrator to hairs in the skin.

Ainsley Iggo stepped in however, to reassure Peter on the basis of work from Mountcastle's laboratory analysis which showed that the vibration sensation with frequencies of up to 250 Hz is detected by Pacinian corpuscles. If this is so, then putting the vibrator on bone is not mistaken because the receptors are extremely sensitive. Ainsley therefore suggested that to test Pacinian corpuscles at say 250 Hz, vibrating a hair would not be appropriate, since it would be limited to a maximum of only about 50 Hz.

George Gordon commented that to find the peak of the tuning curve, when doing vibration tests covering a whole range of frequencies might throw up something new in diagnostic procedures.

The reference to Pacinian corpuscles drew from Yngve a reminiscence from November 1925 when he and Edgar Adrian were making the first recordings from a fibre supplying a Pacinian corpuscle. "For quite a few days Adrian disappeared and from Monday to Friday I didn't see him. So in his absence I thought I could do some experiments on my own and having seen these Pacinian corpuscles in the mesentery of the cat, I managed to get some responses from them. The recordings from Pacinian corpuscles showed that they are so sensitive in the cat that they respond even when you only breath on them very lightly. In fact, Adrian was in training to skate because he was going to Switzerland and he didn't want anyone to watch him. He was very angry when I cycled along and saw him practising on a sewage pond."

Bruce Lynn suggested that it should be possible to differentiate between the response of Pacinian corpuscles and Meissner corpuscles, but in putting forward an idea for doing this he added the proviso that there is always something wrong with a simple idea. He felt sure that Peter Nathan had often tested the vibratory sense by placing a tuning fork against the ankle, and he thought that this must activate Pacinian corpuscles probably in the membrane between the tibia and fibula; it might also stimulate other endings in the neighbourhood but not Meissners. Meissners only appear in the fingers and hands and probably in the toes, but with a tuning fork on a promontory in the lower limb he believed that one might be able to differentiate between the two.

Herbert Hensel then showed us data from recordings in human subjects. In these studies he had compared sensory vibration thresholds with records of activity in single fibres of the human radial nerve (Fig. 4). As can be seen, the sensory thresholds correspond well with the electrophysiological threshold curves in the lower frequency range between 10 and 30 Hz. He considered that the nerve activity set up in this range comes from rapidly adapting mechano-receptors but not from Pacinian corpuscles. At higher frequencies, the thresholds of sensation decrease drastically, whereas the threshold of nerve activity increases and he thought that the vibratory sensation in this frequency range was

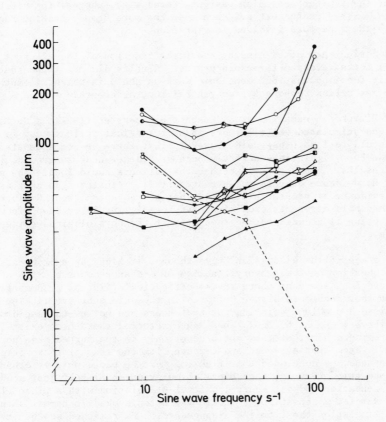

Fig. 4.   Tuning curves for 12 rapidly adapting mechanoreceptors from the superficial branch of the left radial nerve in human subjects (solid lines) and the
human threshold function for the perception of sine wave oscillation applied to
the hairy skin of the hand (dashed line).   A point in the latter curve represents
the mean of 10 subjects, 5 determinations of thresholds were made in each subject.
From Konietzny, F. and Hensel, H. (1977).

probably mediated by impulses from Pacinian corpuscles.

Ainsley Iggo remarked that Herbert's results fitted very well with the analysis made from Pacinian and Meissner corpuscles by Mountcastle, who used local anaesthetic on skin surface to block off the Meissner or hair follicle receptors and found that the flutter sensation was interfered with, whereas the vibratory sense remained. This is further evidence that the more deeply located Pacinian corpuscles are the receptors involved in vibration.

Richard Adrian then asked Ainsley about the function of the afferent C-fibres in the ventral roots referred to in his paper. Ainsley did not think that anyone knew what their function was, but mentioned work on sheep in which attempts were being made to get reflex responses from ventral root afferents.

Manfred Zimmermann recalled an old report by Foerster (1927), a German neurosurgeon who stimulated ventral roots in neurosurgical patients and found that the patients could feel pain but said, however, that there was some debate as to whether he made the stimulation correctly because another neurosurgeon could not get the same results. Yngve asked whether the patients could locate the pain but it seems that no evidence on this had been reported. Ainsley Iggo considered on the basis of evidence at present available, that these ventral root C-fibres are fairly high threshold afferents from alimentary canal, muscle or deep tissues. A nociceptor function is deduced but he thought that the physiological evidence is not yet complete.

At this stage in the discussion Yngve thought it might be appropriate to raise the old question of the sensory function of mechanoreceptive C-fibres, having suggested long ago that there are specific tickle fibres. Commenting on his suggestion that mechanosensitive C-fibres in the skin subserve tickle, Manfred Zimmermann raised a problem, pointing out that there are no low-threshold mechanosensitive C-fibres in human beings. When Yngve retorted that these exist in the cat, Manfred reminded him that they are present only in the furred skin not in the glabrous skin. From that evidence and because tickling sensation can be evoked from the human hand, he concluded that tickling must be based on some other fibres. He had tried very low mechanical stimulation on himself just above threshold for sensation and with very low level stimulation with a piezo-electric crystal, for example, a sensation of tickle was often evoked in normal skin but with heavier stimulation the tickling disappeared. He reminded us that this was an observation which Yngve Zotterman had recorded in his thesis and had proposed that it might be due to inhibition in the CNS. Manfred went on to say that it is quite clear now that when Pacinian corpuscles are stimulated they produce a lot of inhibition in the CNS, and that from electrophysiological work in the cuneate nucleus in the spinal cord and also from psychophysical experiments by Rowe and colleagues (1977) it is established that when the Pacinian corpuscles are stimulated, all other sensations from the skin show elevated thresholds, indicating inhibition. Manfred speculated that tickle is evoked when very few mechanoreceptors with β fibres are excited. A single hair, for example, when moved produces a strong tickling sensation and such tickling sensations occur when single or a few afferent fibres are stimulated and not enough to produce inhibition. So all the afferent channels are without inhibition and therefore there is a very high gain in the CNS for the perception of single impulses. When a stronger stimulation is applied concurrently, a lot of activity is produced in the Pacinian corpuscles which causes inhibition in the CNS. This is afferent inhibition - surround type inhibition and it will reduce the gain of the central apparatus with the result that there is a tactile sensation and not tickle. So tickle occurs in situations when there is no inhibition.

Yngve was not entirely happy with this and referred to a phenomenon which

he always regarded as rather remarkable, that you wake up when a fly lands on
your nose producing only very slight mechanical stimulation. However, Manfred's
interpretation of this is that the stimulus applied is not enough to produce
inhibition and therefore it is perceived as a slight stimulus because it evokes
activity in some mechanoreceptor fibres but not in Pacinian.

Herbert referred back to Foerster's experiments on stimulation of intact
ventral roots. This could mean that he evoked efferent impulses and these might
spread in some way to afferent paths, and this was the conclusion Herbert came to
in his first experiments on human afferent nerves.   The radial nerve was exposed,
a small bundle was split off and cut, to make single-fibre preparations of the
peripheral portion. He said that it was possible to feel the preparation of the
bundle in the receptive field and considered that this could only be explained by
some antidromic or efferent stimulation, spreading to other fields and going up
in the intact nerve.   Foerster observed this when manipulating cut peripheral
nerves. His patients also reported some sensations in the receptive field.

Peter Nathan then made a number of clinical observations. Tickle is no
longer felt after spinothalamic chordotomy, on the side where the spinothalamic
tract has been cut. If you have an intelligent patient and an attempt is made to
find out how much touch goes in the spinothalamic tract and a very light cotton-
wool stimulus is applied, touch is felt on both sides at about the same threshold
but on one side it doesn't tickle whereas on the other side it does. So certainly
in the CNS tickle runs in small fibres. And as further evidence on that, he cited
a very interesting patient who had posterior column lesion, which has the effect
of increasing the stimuli running in the spinothalamic tract, that is to say cold
heat and pain feel more intense on the side on which the posterior column fibres
have been divided. This observation was also made by Foerster (1927). Peter's
patient was very intelligent and when he applied cotton-wool on one side she
localised it by touch and on the other side by tickle; on the increased spinothalamic side she got tickle.

Yngve Zotterman referred to his observations on patients operated by
Sjöqvist for trigeminal neuralgia by cutting the descending spinal trigeminal
tract ("Sjöqvist's tractotomy"), recalling that with cotton-wool stimulation these
patients reported slight touch on both sides but not tickle on the operated side
of the face, just touch. There are no large medullated fibres - only A-delta and
C-fibres in the cut tract. The same holds for the spinothalamic tract in man.
Whatever fibres are concerned, superficial tickle is most likely to be elicited
by the stimulation of specific fibres which give an after-discharge, which so far
has not been recorded from $A\alpha$ and β fibres.   Yngve referred to a letter he had
received from E.D. Adrian in which he wrote "I am prepared to accept your idea of
specific tickle fibres but what is the purpose of specific tickle receptors?".
Yngve's reply was "You should enter a stable and you will see how the horses'
manes and tails are continuously moving - reflexes elicited by the slightest
mechanical stimulation of the skin, developed during millions of years when we,
like other mammals, have lived with and fought the world of insects."

Manfred Zimmermann suggested another aim for the future which occurred to
him listening to the results of self-experimentation and from all the very nice
examples from the experience of patients. He thought that it is clearly very
necessary to increase the neurobiological research in man. This would be
important for the future, not only to save some animals, which would be in keep-
ing with the thoughts of anti-vivisectionists, but also because man is quite a
remarkable experimental animal for which animals are no substitute. He strongly
encouraged more experiments on man, not only neurophysiological but psychophysio-
logical and also neurochemical experiments, mentioning specifically the possibility of taking biopsy material from the brain for tissue culture and to study the

underlying abnormalities in various pathological changes.

Sven Landgren then observed that the discussion had shown that so much is
known about the periphery and so much less is known about central phenomena. The
patient of Sjöqvist that Yngve mentioned could possibly have been influenced by
an injection of dopamine which would entirely change the scene within the
trigeminal nucleus. It could also be changed by sub-conclusive doses of
strychnine. He quoted evidence from Denny Brown who manipulated the receptive
fields of trigeminal dermatomes and showed that sensations could be regained by
these methods and he thought that these findings pointed to important mechanisms
in the CNS which should be taken into account.

In support of what Sven Landgren had said, George Gordon brought up a point
about the cutaneous system which had so far not been mentioned: that is that
every one of the cutaneous nuclei that we know of, is under the most intense
derebrofugal, corticofugal control and what gets transmitted through those pathways may vary under all sorts of concomitant conditions generated in the cerebrum.
He thought that this was something for the future.

Yngve then referred to the observation that when he very slightly touched
the skin on the palm of his hand he experienced a tickling sensation which lasts
for several seconds. There must be an after discharge in the afferent fibres,
and he wondered about sensory fibres responding to light mechanical stimulation
showing an after-discharge. The after sensation is stopped by pressure which
stimulates a lot of large fibres which thus exert their inhibitory influence on
the inflow in the small fibres.

Ainsley Iggo considered that one of the limitations from which we suffer is
that we try to account for everything with what we already know and he felt that
this is a little dangerous because so often the explanations lie in the future.
He thought that because we can use what he described as the marvellous technique
that Yngve was instrumental in giving to us - the single unit technique - for
looking at cutaneous and other peripheral receptors, we then try to analyse all
the central nervous system and the sensory and perceptual problems in terms of
the afferent fibre. He thought that we should not lose sight of the fact that
information gets fed into the CNS and maybe some of the peculiar characteristics
of the afferent fibres are determined by functional limitations imposed on them
by mechanisms which are available in the periphery and this may have nothing to
do with what the CNS actually makes of the information. He suggested that the
future in part lies in a much more detailed analysis in the CNS based on this
wealth of information we now have of the properties of peripheral receptors, but
not expecting that we can simply analyse the sensation in terms of the peripheral
afferent discharge.

Carl Pfaffmann added a final comment to the discussion on the central
mechanisms and their relation to the sensory apparatus and the study of clinical
conditions. He wanted to make a plea for analysis by precise methods of animal
psychophysics. These methods are quite well developed. He reminded us of the
examples of that given by Dan Kenshalo and said that in other sensory fields it
is quite possible almost to match the verbal report by well controlled animal
psychophysics on the same preparation, the same sense organ and the same system
on which physiological information, peripheral and central is available.

REFERENCES

Ferrington, D.G., Nail, B.S. and Rowe, M. (1977). Human tactile detection thresholds: modification by inputs from specific tactile receptor classes. J. Physiol. 272, 415 - 433.

Foerster, O. (1927). Die Leitungsbahnen des Schmerzgefühls und die chirurgische Behandlung der Schmerzzustände. Urban and Schwarzenberg, Berlin, 360 pp.

Gross, G.W. (1973). The effect of temperature on the rapid axoplasmic transport in C-fibers. Brain Res. 56, 359 - 363.

Gross, G.W. and Beidler, L.M. (1975). A quantitative analysis of isotope concentration profiles and rapid transport velocities in the C-fibers of the garfish olfactory nerve. J. Neurobiol. 6, 213 - 232.

Konietzny, F. and Hensel, H. (1977). Response of rapidly and slowly adapting mechanoreceptors and vibratory sensitivity in human hairy skin. Pflügers Arch. ges. Physiol. 368, 39 - 44.

Ochs, S. (1972). Rate of fast axoplasmic transport in mammalian nerve fibres. J. Physiol. 227, 627 - 645.

Pease, D.C. and Quilliam, T.A. (1957). Electron microscopy of the Pacinian corpuscle. J. biophys. biochem. Cytol. 3, 331 - 342.

Quilliam, T.A. (1966). Unit design and array patterns in receptor organs. In: A.V.S. De Reuck and J. Knight (Eds.), Touch, Heat and Pain, J. & A. Churchill, London, pp. 86 - 112.

Quilliam, T.A. (1975). Some hazards in the interpretation of the pattern of structure in lamellated receptors. In: The Somatosensory System, Ed. H.H. Kornhuber, pp. 200 - 203. Georg Thieme, Stuttgart.

Raderman-Little, R. (1979). The effect of temperature on the turnover of taste bud cells in catfish. Cell Tissue Kinet, 12, 269 - 280.

Santini, J. (1969). New fibers of sympathetic nature in the inner core region of Pacinian corpuscles. Brain Res. 16, 535 - 538.